The War We Won

✯ ✯ ✯

BY L W SWARTZ

DORRANCE
PUBLISHING CO
EST. 1920
PITTSBURGH, PENNSYLVANIA 15238

Dorrance Publishing Co
585 Alpha Drive
Pittsburgh, PA 15238
Visit our website at *www.dorrancebookstore.com*

ISBN: 978-1-6366-1485-4
eISBN: 978-1-6366-1662-9

Preface

Max and James, I am writing this memoir so that you boys might understand that what I did—joining the Army—was not an idle adventure. I also want you to understand that ours—all those who fought, bled and died—was not a failed effort. We who went into combat won our part of the Viet Nam War. It was our leaders, egos fluttering in the self-adulating winds of Washington, D.C., who threw away all that effort and sacrifice. I hope that this little trifle will help you to understand what I experienced and what it was like. I regret that my prose is not nearly artful enough to fully convey all that my friends and I went through. If you get nothing else from this know this: They were good men, boys really, who went to do what their leaders deemed the necessary thing, and the failure to secure victory is the ultimate betrayal of their sacrifice by those same leaders. Do not ever allow yourselves to be lured by those who consider themselves the best and the brightest because the only service they provide is to themselves.

Chapter One
Going In

On the twentieth anniversary of my birth, I took the oath as a soldier in the United States Army.

That action had been sixteen months in development. I had been thinking about enlisting since my senior year in high school. The army had a program that allowed individuals without a college degree to enlist and become pilots. All I had ever wanted was to fly, and I didn't see the need for a college degree to accomplish that goal. My mother had much to say about that attitude, none of it affirmative.

My mom worshiped at the altar of education. It was at her vehement insistence that I enrolled at Northern Illinois University. In high school, there had been some conversation and activity regarding the conflict raging in Viet Nam. I had gotten into verbal spars with classmates over my belief in the war and why we were fighting. None of it was too volatile. College was a whole new world of animus. I thought I had met ardent antiwar arguments when I would have discussions with my high school classmates, but none of my previous experience prepared me for the encounters that filled the nine months from September 1967 to April 1968.

I was not allowed to support the war and remain in the human race. No argument I could present, no atrocity I could cite was acceptable as support for my position. A girl I met in October, Linda, was even drawn into the fray when her roommate's boyfriend found out I was a supporter of the war. When it was brought to light that I was a "war monger," Joyce went to war with harangues against my character and Linda's by association. There were multiple weekends when Linda was forced to retreat to the sanctuary of her home because things had become so overheated in the dorm. Even the minister that served the college campus delivered sermons against the war.

The comment that rang home most was when I had been in a verbal spat with a protestor. "Easy for you to be for the war while you're sitting here enjoying the good life. If you feel so strong about the war, you should put your ass on the line!" That statement hit home with me. I thought about it a lot.

Academics were not my forte. That fact combined with the writhing hostilities on campus made my freshman year far short of a success. In May 1968 I got my grades from Northern. It was the confirmation that I needed. While my mom and sisters were on vacation to Montreal, I went into the Army Recruiter in Mt. Prospect. I arranged to take the exam to determine whether I was eligible to go to flight school, and in the broader view, whether I was physically eligible to enlist. I took a day off from my summer job and went in to Chicago to take the preliminary exam.

When I got to the induction center, I was seated in a large room. The walls were in need of a fresh paint job and litter clung to the corners of the room. Paint was peeling from the exposed ductwork. When your function is the intake of draftees, it was not apparently necessary to put on a neat façade for your clientele. It wasn't like they could decline enlistment on the basis of substandard intake facilities. Not that any of those accompanying me in this adventure were paying any attention to their surroundings. They could be seated in the atrium of Waldorf Astoria and their ultimate destination would be the same.

There were seventy or so young men seated with me in standard school chairs with attached writing surfaces. Everyone but me was there having received their draft notice. They sat there awaiting their placement. I had been sitting, listening to various conversations between groups of young draftees, when a Marine stepped into the doorway. He looked like he had stepped out of a poster in a Marine recruiter's office.

"I need five volunteers for the United States Marine Corps," he announced in a full baritone voice. He easily moved from an attention status to what I would learn in basic training was an at ease posture. Hands withdrawn behind his back he surveyed the room of draftees.

The response in the room was universal. Bodies shrank, and heads turned to avoid eye contact as men confirmed to each other that they wanted no part of the Marines. After five uncomfortable minutes had passed, a short pimply faced private wearing a tan Army class A shirt and green slacks with the shirt partially untucked stepped into the door way. He looked around the room and then up at the Marine standing next to him.

"Get your volunteers, Sarge?"

The Marine never broke his gaze. "Not yet."

The private grinned and then looked across the room. I wondered what I would do if the fickle finger of fate selected me. His right hand rose and began pointing at various men in the room.

"You, you, you, you, and you all just volunteered for the United States Marines. Congratulations." He turned and started to leave the doorway but stopped short. "There you go, Sarge." Then he disappeared from sight. The five victims stood and moved to the doorway, mumbling to themselves at the fates that had just compounded their misery, and I was spared the problem of explaining that I wasn't a draftee to anyone.

I was finally ushered to an exam room where a young doctor began a physical examination. I was asked to bend and stretch. My blood pressure and pulse were taken. I was given an eye exam, and he checked my hearing. I was weighed, and my extremities were inspected. He settled on my right hand.

"Are you right handed?"

"Yes sir," I responded.

"How long have those fingers been like that?" He was fondling the index and middle fingers of my right hand.

"Since I was twelve."

"How'd it happen?"

"Got them caught in a hand crank corn sheller."

"That had to hurt." He felt the joints and checked the color of the digits. "Can you shoot a gun with these? Because if you can't, you are 4-F."

"Sure can, I've been able to since they healed."

He looked at the fingers skeptically. "I can make you 4-F no problem."

"No, sir, they're fine." The thought of being 4-F had never entered my mind, and actually, as it was being proposed, the thought was repugnant to me.

"Come to attention and give me a salute."

I stood at attention in my skivvies and drew my right hand to the corner of my right eye in what I considered my best military salute. The doctor moved in and adjusted my right hand to what he deemed a proper salute. He stepped back and moved from side to side, focusing on my hand and the salute I was attempting.

"I'll be right back." He sat his clipboard down on the exam table and left the room. I allowed the salute to drop and stood at ease. I was becoming aware of the cool temperature in the room so I placed my hands under my armpits. In several minutes he returned with another doctor. I dropped my hands to my sides. I had to smile to myself over a resisted urge to smell my hands.

"Give me that salute again," the first doctor ordered.

I raised my right hand to the corner of my eye again as the first doctor had shown me. The two doctors took turns looking from various angles at my right hand. I felt like a secondhand piece of art being evaluated for value and authenticity.

"In that salute you really can't see it," the new doctor offered. He then looked at me. "You can shoot a rifle with those?" He fondled the two offending fingers and tested the joint. "Pretty stiff," he murmured.

"Yes sir," I responded quickly.

"You're enlisting?" he asked half in disbelief. I nodded my assent. "I'd say he's good to go." The new doctor turned and left the room.

"Okay, son, get your clothes on. Someone will be in to take you for the rest of your testing," the doctor who had examined me said. He picked up the clipboard and began making notes on the sheet of paper held by the huge metal clamp.

I got my trousers and my shirt back on. Once my clothes were back on, I sat in a folding chair that sat in the corner opposite the exam table. I got my shoes and socks on. I was finishing with the left shoe when an army clerk came in.

"Come with me, you're going up to the second floor to take the written test."

The written test reminded me of something they give to analyze an individual after a stroke or a brain injury. Multiple boxes containing crude pictures were displayed on each page. Each box depicting the view through the wind screen of a helicopter. You were challenged, from the picture, to determine whether the aircraft was in a climb, a dive level, or banking right or left. I went through the test and turned it in. I was told that they would advise my recruiter how I had done. I left the induction center uncertain as to where I stood.

As I drove home, I was struck by a sudden fear: What if they called me after my mom got home? I hadn't said a word to her about this, and I had no intention to until I knew I was in and set to be in the flight program. My hope was that the army would act quickly to get me inducted and my scores would be high enough to qualify me for flight school. I didn't need to fear. The recruiter was on the phone that evening. Everything was good. The only issue now was when did I want to go?

Linda, the girl I had met at Northern, the only positive to the nine months there had continued to date me after spring semester ended. When I was sure I was going to enlist I, had made an attempt to break up with her. It had been half-hearted at best. After a forty minute conversation on the patio in front of the student union, I ended up confirming our relationship instead and that I wanted that to continue for the summer, so I set my enlistment for the fall.

I had one last hurdle to overcome. My mother had to be told of my action. The thought had been briefly entertained that I could just get up on the morning, pin a note to the bed that I would be back in three years, and take off for the army. It was a coward's approach. I knew I couldn't really take that route, but the relief in my fear was comforting. As has always been my want, I procrastinated until the end of July to break the news to her. I broke the news on a Saturday as we were preparing for a party in the backyard. My thinking had been that with guests coming, there would be no opportunity for a prolonged argument.

She took the news rather stoically. That unsettled me more than the expected rant would have. Her only comment was "we'll see." That was a gauntlet thrown down that few had the nerve to pick up. Unfortunately, I had picked it up more than a month earlier, and there could be no dropping it now.

At one point during the party, as my Uncle Don was cooking the brats and hamburgers, my mother came over, and loud enough for me to hear said, "Before you leave I need to talk to you about your nephew." My uncle looked at me and asked what I had done that he had inherited ownership of me. I shrugged and said that he would be hearing all about in a little while.

After the cleanup from the party, when the grill had been moved, the folding tables put away, and the trash collected, we all collected in the English basement of our house. The basement had a large picture window that looked out over the backyard. There was my mom, my uncle and aunt, Linda, and me. My mother started off with the indictment.

"Do you know what your nephew has done?" There was not a large enough gap for my uncle to reply. She immediately revealed my sin. "He went and enlisted in the army." To emphasize the enormity of my sin, my familial alignment had been shifted to my uncle. I was no longer her son; I was Uncle Don's nephew. She didn't wait for his response. She demanded that my uncle tell me I couldn't do that. My uncle looked dumbstruck; how was he supposed to respond to that command?

Finally he said, "Barbara, I can't tell him he can't do something that is clearly legal for him to do."

There was a pause. The response she had just received was not the one she had expected. I could see her marshaling her next argument. When he said that that was not possible either, she took to the issue of my fingers. Confident that I had not disclosed the old injury, she announce that she would call the army and let them know that I was crippled and unable to perform the necessary functions of a soldier. My uncle and I addressed all of her objections as clearly and as calmly as we could. It is a truth of life that logic and calm rarely counteract emotions, especially the emotion of a parent wanting to protect their child from potential harm, even when that potential harm is self-inflicted. As I had feared, the evening ended in tears and recrimination. My uncle, clearly shaken by the confrontation, shook my hand and offered his hope that I was not making a huge mistake.

There was a weeklong period of cool tolerance between my mom and me. The anger gave way to fear and concern. At the beginning of August, she even began addressing what I would need when I left for basic training. I got my report date, and I could not believe the date the army had chosen. Mom's response was to ask if the date couldn't be pushed back to allow me to celebrate my birthday with my family. I assured her that was not possible. Her only comment was that this was another way that my decision had ruined things. The weekend before I went in, my Uncle Don took Linda, my mom, and me out to an elegant dinner. He even conned the waiter into serving me a Scotch on the basis that I was going into the army. After some sweet talk, a Scotch was brought, and I had my first legit drink.

The summer had passed with lightning speed. Far too quickly I stood in the Military Induction Center in downtown Chicago having taken an oath to defend the Constitution. Our orders had been prepared for a group of four of us to be shipped to Fort Polk, Louisiana for basic training. I was placed in charge of the group with all of the paper work being handed to me. I am certain this was because I was the oldest of the group at the ripe old age of twenty, or perhaps it was a birthday present. Regardless of the army's reasoning, I was given our orders. The four of us were packed into a van, and we were taken to O'Hare Airport. United Airlines, ever efficient in their provision of air transportation, had us aboard the plane and in the air on time. As the plane, a 727, climbed up from the runway and began a bank to the west, I watched my world, the one I had resided securely in, slip away beneath me, a blur of lights stretching out toward the western horizon.

In Dallas we collected our baggage from the United luggage claim area and started our search for the connector to Fort Polk. After leading the group in the wrong direction for several minutes, I stopped and asked a sky cap where Trans Texas Airlines was.

"Tree Top?" He smiled. "That's the other way." He elevated an arm to reinforce the verbal direction.

'Tree Top?' I thought, that doesn't sound promising. We turned about face and headed the other direction. One of my charges muttered something about knowing the right way all the time. I chose to ignore the comment.

At the far end of an isolated terminal, we found the counter for Trans Texas Airlines. A middle-aged gentleman wearing a gray vest, white shirt, and stained tie looked up at our approach. He smiled a tired smile, indifferent in its application.

"How may we help you, ah, boys?"

"We are supposed to catch a flight to Fort Polk." One of my cadre usurped my authority with their out of turn comment. I looked back at those behind me trying on my assertive face.

"I have our paper work," I offered, attempting to regain my proper role as temporary commander.

The man took the papers I had extended. "Thank you, general," he said in an indifferent voice that carried a Southern accent. Several of the guys behind me sniggered, and I could feel my face turning red. It didn't matter, I tried to tell myself. The clerk finished processing the papers and returned them to me. "Have seat over there," he said without a gesture or physical indication of where he meant us to sit. "I'll call you when the plane is ready to board."

We looked around and spotted a row of seats along an empty wall. There were two women seated and two children crawling along the tiled floor in front of them. The women had taken root right in the middle of the row. We looked at each other, and in quiet agreement we all opted to stand. There wasn't long to wait. The clerk who had processed our papers came around the counter and announced our plane was ready to board, and he asked us to follow him. There was a door some twenty feet down the hallway where he stopped. He took hold of the doorknob and pushed. A rush of air and noise poured into the hallway. The children backed into their mothers' legs and took a firm grip on the shelter they promised. The mothers bent over and hands on shoulders directed them ahead into the noisy night air.

The plane was a vintage DC-3. A thick blue stripe ran over a row of windows. Over the strip were the large yellow letters TTA. A short stairway was braced against the fuselage at the doorway. We all allowed the women and their kids to board, and then it was our turn. We moved up the aisle, literally as the plane was a tail dragger and the aisle sloped up toward the cockpit. I took a seat next to a window and looked out at the sprawling complex that was Love Field. Jets of every description rolled along the cement tarmacs. Red and green lights strobed like night fireflies looking for a mate. Off in the distance, the bright white lights of a jet on final approach shone like a star.

My attitude during the summer could only be described as a smug one. Once I had crossed the hurdle of telling my mom of my enlistment, a great hubris seized me. I was at last on the path I knew was mine. I would enter a career I had wanted since my eleventh summer in Waco, Texas. My uncle was an Air Force pilot, and the vision of him taxiing out of a thick fog and his plane magically forming out of the summer vapor had mesmerized me as the jet engine whined to a stop. The cockpit cover magically raised revealing my uncle. He had his helmet under his left arm. An airman rushed a ladder to the side of the plane. He was talking to another pilot who had landed with him. They were talking about the flight and laughing at the weather. In the phantasm of that moment, I was seized by the desire to be a pilot.

Now that goal fermenting for nine long years was to come to fruition, I was going to be a pilot. The smug adjective grew less apt as my report date approached. The nearer my date with basic came and my proximity to Ft. Polk got closer, my attitude turned to concern. What would basic really be like? Would I be able to measure up? Could I really fly, and more importantly, what would happen if I didn't become a pilot? What ever could I be if not a pilot? The thought was too grim to contemplate. The desire to be a pilot was too real and too close not to come true.

The two pilots passed up the aisle and took their place in the cockpit. Their presence snapped me back to the moment. I watched them go through their checklist and flip switches and turn dials. I was absolutely fascinated by the whole process. An older woman announced that we were prepared for takeoff and asked everyone to please buckle up. The engines fired off, slowly revved up their RPMs, and then backed off. Soon enough we were rolling along the tarmac and then taking position at the end of the runway.

Once again the engines revved up, and the plane started its roll. The aisle rose to level and then again dipped as the pilot put the plane into a shallow climb. We banked, and I watched civilian life slip away beneath the wings. Dark emptiness glided below us as my ears popped several times. Looking down for some visible signs of the ground below, there were none. Unlike our departure from Chicago, Texas had a distinctly barren look in the dark. Only the occasional blot of light flickered across the black landscape. Tree top airlines, I didn't think so; we were clearly well above tree top level. How much above tree top was unclear. The sound of the two radial engines droning along, pulling the plane through the air filled the cabin. I looked around the cabin. Everyone was as quiet as a Sunday congregation.

We had been airborne for all too short a time when I noticed the plane making a series of slow turns to the left and right in an indifferent pattern. I could see a bright light out my window. It slowly disappeared, and then we made a sudden turn back toward it. I heard one of the other guys wonder aloud whether we were lost. The engines throttled back, and we began a slow decent. I could hear the hydraulic sound of wheels coming down and the flaps being lowered. All the distinctly mechanical sounds a plane makes as it prepares to transition from flight to ground crawling.

The plane reluctantly touched down and then rolled to a stop. The right engine revved up, and the plane pivoted to the left. We rolled to a stop under a bank of bright lights. The hatch was pressed open by the stewardess. The same humid air we had left in Dallas pushed into the compartment. We allowed the women and their kids to deplane first, and then we all, in turn, rose and moved to the rear. The stewardess wished us each a "good luck" as we passed out the door. Two men stood across a short span of cement under a pair of lights. They both sported sergeant stripes on their perfectly creased sleeves. Their campaign hats sat perfectly square two fingers above their brow lines.

When we appeared, they moved from an at-ease to an attention posture.

"Welcome to Ft. Polk. I am Sergeant Kinchen, and this is Sergeant Hammerly. You people will be under our tender care for the next eight weeks while we attempt to make men of you sorry piles of crap. Get your bags and follow us." When we were slow to respond, we were greeted with a thunderous "MOVE IT." We quickly retrieved our bags and followed the two sergeants. We were lead to a large building clearly designed for the processing of large numbers of men.

There was a large group of men already in the building. Once we were all together, we were formed into lines. The first thing was a review of the contents of our bags for contraband.

"What's this?"

"My meds for acne, sir." I winced at the faux pas. I knew better than to call an NCO, sir but there it was. I didn't have to wait long for the response.

"I am a sergeant. I work for a living. I do not live off the fat of the land. You will address me as Drill Sergeant, not sir."

"Yes."

"Yes what?"

"Yes, Drill Sergeant."

"Now what is this?" He held up the plastic bottle of pills I was taking to keep my acne in check.

"Acne medicine."

"You have a prescription for this?"

"Not on me. I could have one sent if you'd like," I volunteered.

"No. That's not necessary. You know what causes acne?" He didn't wait for my surmise but plunged into his diagnosis. "Corrupt living, too much candy, too much dope, or too much pussy. Looking at you,

I'd say it's too much candy cause you don't look like you ever got any, and you look too afraid to do dope." He took the bottle of pills and threw them into a large metal garbage can. "We're going to clean you out so you don't need those pills. You're going to be treated for those pimples the army way."

Once he finished going through my bag, I was dismissed to a line of other recruits who has been purified. When everyone had gone through the same process of being stripped of any contraband, we were marched to another building where the most personal of all property was stripped from us. Several guys had shoulder length hair that joined the growing pile of keratin on the floor. When each of these men was separated from their long locks, one of the drill sergeants would comment, "Well lookie here. There's a maggot under those pretty curls."

After being separated from our hair, we were shuffled to another building where we were issued military clothing. Boots, socks, boxer shorts, pants, blouses, shoes, hats, and a duffle bag were piled up into open arms. We were not given any time to organize anything. We were run to the company area, my home for the next eight weeks. Sgt. Kia stood in front of us as we stood panting and struggling to keep all of our newly acquired clothing from falling on the ground.

"Welcome Company E, Third Battalion, First Training Brigade. We are going to make men of you people over the next eight weeks. That process will begin tomorrow. Tonight we will be getting you into your barracks."

One barrack was filled with all guys from Iowa. Our barrack had several National Guard recruits. We were soon to discover that while we were all lower than whale shit in the hierarchy of life forms that occupy the planet. our first duty as fire guard was assigned, and the day ended with bunks made and sleep fitfully achieved.

Chapter Two
NOTHING BASIC ABOUT IT

Basic training began at 5:30 the next day. The clatter of a garbage can bouncing down the center aisle followed by the drill sergeant's baritone voice announcing that all feet had better be on the floor before he reached the end of the aisle. The thump of feet hitting the floor filled the open room. We all rushed to get into the latrine. Everything was open. Toilet stalls and sinks all sat in open rows. Men rushed to take care of their necessaries. Men rushed to get shaved, managed to nick their faces, and toilet paper blotched the faces of half of the guys in the barracks.

Dressed and cleaned, we collected in the area in front of the three two-story buildings that would be home. The milling quickly stopped with the appearance of Sgt. Kia. He was a slender man with a weathered face and an indistinguishable accent. His mien was all business, and there was the presence that told all with eyes and ears that he was all business and not to be trifled with.

"Fall in!" Men began to collect in lines. Once we were in the best order we could manage, we were shown how to properly line up. How to dress right, your right hand on your right hip elbow pointed at a ninety degree angle touching the man to your right on his left arm. Once we were properly aligned, Sgt. Kia introduced himself and the other tormentors we were to come to know and love.

I had watched all of the war movies, especially *Sands of Iwo Jima* so I had some inkling of what the basic training experience would be like. Having an idea of what was to come and actually experiencing it were two distinctly different things. Learning the commands given to a group of men was one of the first lessons. After we properly formed, we were marched to the mess hall for our first meal as a unit.

It wasn't that the food was of poor quality. I am certain that when it arrived, it was as good as any institutional food can be. It was the brutality of its preparation that gave it that certain flare. Runny scrambled eggs, shredded potatoes that were pronounced hash browns but were in reality hash blacks. Sausage that Jimmy Dean would retch at and all pigs everywhere would be ashamed to have to admit they had died to produce. There was a clear philosophy at work where food prep was concerned. It doesn't have to taste good; work them to death and let hunger take care of the rest.

After our breakfast was finished, we were marched to a classroom where we were introduced to all things military. Ranks were discussed and the proper honoring of officers and noncoms. After classroom came physical exercise. The lumps that the draft boards had given to this cadre needed to be molded into fighting machines. We were called by name and asked for our serial number. Each number had a prefix RA NG etc. When one of the National Guard recruits would yell out his NG and number, one of the drill sergeants would respond, "I'd rather have a sister in the whore house than a brother in the National Guard."

It was clear from the conversations I overheard that the National Guard guys were in it to avoid the draft and a real future in combat. One guy was the captain of the Cornell track team; another was a jock from Pennsylvania. For all of that, the first few days of PT were every bit as much a struggle for them as they were for the rest of us. Pushups, jumping jacks, leg lifts all done together, and all done until you were certain that your body could not possibly do one more repetition. But the fear of drawing attention from the ever watchful eyes of the training sergeants made us keep the effort up despite pain and exhaustion. We were issued dog tags with our name, service number, blood type, and religious affiliation.

The first lesson I learned without formal training was that regardless of your efforts, no inspection could be passed. The keen eyes of the drill sergeants could find a microscopic mote somewhere regardless of your efforts. They could examine an amoeba sans microscope. Every dirty discovery was pronounced a blight upon humanity. I had cleaned my cubicle to within an inch of its life only to have the drill sergeant take a finger and run it along a curved piece of metal that formed the top of the doorframe. Small motes of dust filtered down into the sunlight.

"Are you a pig? It looks like it to me. You must like living in the mud and filth. My mother would faint dead away if she saw this mess." It was clear that to fail was to fail on a grand scale. There were no small failures. Cleanliness might be next to godliness, but in the army, it was next to impossible.

One of the NGs, the track star, noticed after several weeks that he had failed to evince an erection since the beginning of basic. His thesis was that it was true they were putting saltpeter in the food. He started making inquiries with the guys who had done KP. He asked each of them if they had seen the cooks doctoring the food. When these queries were unfulfilling in supporting his theory, he went straight to the source.

The cooks laughed. "No man, we don't put no saltpeter in your food. Maybe your equipment is just on the fritz." This was clearly not the thing he wanted to hear. His concern became so magnified that soon everyone in the cadre began calling him "saltpeter." To the day of graduation, he persisted in his firm belief that our food was being treated.

He advised any who would listen that his first project upon returning home would be to cleanse himself of the chemicals that had reduced his manhood to nothing more than a water hose. His greatest fear was the uncertainty of how long that cleansing process would take. Several of his fellow National Guard trainees assured him that he would be up and running as soon as he was in his girlfriend's loving embrace.

We were given aptitude tests to find out where we might best be placed. After my exam, I was advised that I had an aptitude for language and I might want to change my MOS. As the sergeant put it, a second language skill would be very useful out in the world. I found this hugely entertaining, and I was certain that my high school German teacher would find it equally amusing. Knowing my struggles attempting to learn the German language, I told him I would pass; I wanted to be a helicopter pilot, and that was it. He acknowledges that there was a binding agreement in place, and there was nothing he could do to dissuade me from my goal.

We were given classroom and practical training in a variety of areas. Orienteering was one that I found especially informative. And by some happenstance, I seemed to have some skill at it since I was one of the ones who could successfully navigate a course. Physical training was a daily event. Even when we were in the classroom, they would weave physical training into it by having us run in formation from one place to the next.

We finally were issued our rifles: the M-14 firing a 7.62 millimeter NATO round. We were set about the task of learning to break the rifle down and put it together. We cleaned our weapons and then had them inspected by the drill sergeants. This was where the fact that there was no such thing as clean was reinforced. I was sure my rifle was as clean as a weapon could ever be made only to have my drill sergeant find a minuscule speck on the front sight. I would have defied anyone not equipped with a microscope to see that dot, but there it was. The drill sergeant's judgment was not open to debate, so I broke the rifle down and cleaned it again, paying special attention to the sights and the barrel. My second attempt was met with a nod of approval. I felt as if I had just won a marathon.

The pugil pit was not my favorite. The pugil sticks were long dowels of about six feet with padding on each end. We wore pads to protect our heads, but the object was to simulate bayonet fighting. In reality it was an opportunity for the drill sergeants to watch us beat each other into pudding. They would try to match men to others of similar size and weight, but that didn't always work out. And there were a few they chose to punish for infractions by matching them with others of greater size, weight, and reach. While the two combatants set about striking each other with as much ferocity as could be mustered, the rest of us stood around the combat circle and were encouraged to chant "Kill, Kill, Kill." In the heat and humidity of central Louisiana, it didn't take any time at all to be soaked with sweat.

In truth the goal of most of us was to go through pugil combat with a minimum of injury and to have earned some modicum of respect from the drill sergeants and more importantly our fellow trainees. No one wished to be seen as a loser in the field of armed combat. We were here to learn to be instruments of war. To be deemed ineffectual in this avenue of endeavor was not acceptable. That was true with regard to both our fellow troops and our drill sergeants. As a unit the opinions of the members carried a lot of weight.

Those who approached the task with insufficient enthusiasm were accused of not having a pair, and they were asked if they some kind of girl. After a several minute diatribe from one or more of the drill sergeants, even the most recalcitrant of participants entered the circle with newfound enthusiasm. Once they got with the program, they would hear words of encouragement.

Honey Bear was one of those they bent every effort to inspire. He was a kid who was grossly over weight. He was also extraordinarily hairy. I had seen men stripped to the waist who were well covered, but nothing came close to the pelt that Honey Bear carried. He could not keep up during runs, and he was always falling out during marches. The Louisiana climate was made worse by the furry mat that covered his torso. Despite the restricted diet and exercise regimen, he refused to lose weight. It was somewhere in the fifth week that he was cycled into a special training platoon to see if the army could not get him into shape. The five weeks he spent with us were not something that bore a lot of watching. One or the other of the drill sergeants was always on him to move out with a purpose or move, move, move. I, nor any of my fellow maggots, wanted any part of what he went through.

Honey Bear was the one who ruined our one hundred percent participation in blood donation. Try as they might, the medics could not find a vein to draw blood from.   Sheaths of fat sheltered every prospective vein. There were two things that every training company strove to get one hundred percent participation in. Blood donation was one; buying government bonds was the other. Sergeant Kia announced that they were going to sign us up for the automatic deduction from our pay.

"I cannot make you buy a bond! But, by God, you will buy a bond," was how he put it. As with Honey Bear and the blood donation, there were a few who chose not to go along with the program. After a day on the obstacle course with special attention from several drill sergeants who behaved as if keeping their rank depended on it, the three free spirits were more than happy to sign up for a bond. Sgt. Kia had his one hundred percent participation in this case. He had made good on his oath that we would all participate and no one had been struck.

We learned how to throw a hand grenade. Just like in the movies, one of the trainees lost control of his grenade. The call went out, and everyone dove for cover. Just like in the movies, no one was hurt, but the trainee who lost control of his grenade never made that error again. The drill sergeants had a way of driving certain points home. I had remembered an incident in one movie where a trainee allowed his rifle to fall. The resulting punishment was a required run around the company as everyone else marched. Not wishing to tempt fate even a little bit when we would stop, I always kept my rifle butt down between my legs.

On our second march back from the rifle range, we stopped for a ten-minute rest. Sure enough, as soon as we were all settled, I heard the distinct sound of a rifle falling. In unison that any coach of a pro team would be proud of, the five drill sergeants yelled out, "Who dropped their weapon? Who was the careless maggot who just jeopardized his and his buddies life?"

When the feckless individual identified himself, the five drill sergeants descended upon him, like a scene from the nature channel where ravenous piranhas roil the water around their victim until only blood and body part remain. The five men surrounded the soldier and in unison degraded him verbally. One of the most important tenets of the infantryman is the care of that instrument which makes him a useful weapon. The rifle is all important, and its care is paramount. True to the scene from the movie, the guy ended up running around the rest of the company as we marched back to our barracks. He held his rifle in the air, and he chanted his love for his weapon and the callous deed that had disgraced him and his unit. The point had been made, and disregard for the care of our rifles never occurred again.

We were trained in the use of gas masks. After instruction on its proper deployment, we were led, in small groups, into a bunker. We were ordered to put on our gas masks, and once we were all properly set, they discharged tear gas into the bunker. We sat there as the drill instructors walked up and down the row watching us all for any reaction. After what seemed an eternity, we were instructed to stand up and prepare to exit the bunker; as we headed for the exit, each of us had our masks pulled off. We were exposed to the tear gas for just a moment, but for the first ones that was long enough. I saw what had happened to the front of that line, so as I approached the drill sergeant, I took a breath and closed my eyes. I managed to avoid the worst of the exposure.

In addition to using the M-14, we were taken to a range and handed BB guns. Empty shell casings were lined up on a wooden board; these were our targets. We were staged ten feet from the boards. Each of us was tasked with shooting the shell casings with the BB gun. The point was to not aim, in the traditional sense, but to shoot with speed from the hip. This proved to be reasonably entertaining for all of us. I imagined myself as the rifleman pumping rounds into my foe. I neither excelled nor failed miserably at this exercise. I had to resist the temptation to draw my BB gun up to my cheek and take true aim at the targets. Once when I had forgotten myself and started to lift the weapon up, I heard an "Ahahah." I looked over my shoulder to see the instructor watching me.

We also were taken to a pistol range where we were shown the use of the 1911 forty-five caliber semi-automatic pistol. After being instructed in the proper grip and how to load and charge the weapon, we were placed on the firing line. Each of us had seven rounds to shoot at a target that was twenty-five feet down range. I fell in love with this weapon and vowed to myself that this was a weapon I would carry into combat. I managed to shoot better than most. All seven of my shots hit the target, three fell tightly in the first circle, and one was in the bull's-eye.

All of us had to serve KP and walk fire guard. Fire guard was a detail that required the men on duty to walk the perimeter either inside or outside of the barracks. If a fire was detected, the man on guard would rouse all of the men asleep to get them out of the building. The barracks were old wooden, two-story structures that had been built during World War II. They could go up in flames quickly, and the fire guards were meant to be the safety barrier.

One early morning, I was walking fire guard outside the building. As I passed the open window of one of the sergeant's rooms on the first floor, I heard his radio. It was tuned to WLS the Big 89 in Chicago. I stopped under the window to hear the music that was filtering through the screen. I felt like I had been whisked home. Larry Lujack, Super Jock, was doing his DJ duties. He introduced each song and added dashes of sarcasm as only the kindly and delightful ole Uncle Lar could. After a song, I started my round. I discovered that my pace was significantly brisker when I got out of earshot, and it slowed to a crawl when I could hear the radio again. My mood for the rest of the day was totally elevated. I didn't even feel any pain during the morning one-mile run.

Bivouac was another adventure. We hiked out to a camping area. We set up individual tents and prepped them for rain with a trench around the outside of the tent. The bivouac was a military operation, and guards were posted as if we were in a combat situation. A lieutenant took the notion that it would be fun to infiltrate our perimeter. He got into the perimeter and tossed a teargas grenade into one of the tents. Unfortunately for him, his penetration did not go unnoticed. As he set about exiting the location, one of the guards waited behind a tree in his path. As he got even with the guard's position, the guard let him have it with the butt of his rifle across the lieutenant's forehead. The lieutenant went down screaming about being assaulted.

Sgt. Kia was summoned, and the lieutenant went into a rage demanding that the guard be punished. Sgt. Kia stood listening to this tirade for a long period as the lieutenant's voice grew more nasal as the swelling began. When the lieutenant had exhausted his fur,y Kia said there would be no punishment. The man on guard was fulfilling his duties as instructed, and the lieutenant had been an unidentified intruder. In point of fact, he had not even advised the officer-in-charge of his intensions.

Nothing came of the event, and our esteem for Sgt. Kia grew again. There would never be an absence of fear, but we all had a firm feeling that as long as we toed the mark, he would have our back. There was a comfort in this knowledge that made all the rest of basic more than just tolerable.

Night infiltration came toward the end of training. We were taken out to the night infiltration course. We had been advised that the explosions and bullets would all be real. We would enter a trench at the base of the course. On orders we would go over the top and low crawl through the obstacle course. The low crawl is a method of advancement where knees and elbows are employed to propel oneself forward exposing a minimum of the body to enemy fire.

One of the instructors assured us that the machine guns were mounted so that they could not shoot lower than four feet over the course surface. That was okay if we stayed in a low crawl, but if we stood up, there would be problems. We entered the trench at the end of the course. There was a slow incline that ran the full length of the trench. To our right were target silhouettes posted on boards. Once we were all in the trench, the machine gun fire began. As I looked up I could follow the trajectory of several tracers as they passed neatly through the target silhouettes. I looked back at the man behind me. I could see he was watching the same thing I was.

"That ain't friggin' four feet high."

We looked around for someone to express our concern to. There was no one. At that exact moment of despair, the order was given to go over the top. Against my better judgment I went over the top and began to crawl toward the other end of the course. I was next to a sandbagged crater when a huge *hawumpf* erupted sending dirt and smoke out in every direction. I had been getting tired, but the blast gave me new inspiration, and I found myself passing several of my comrades in their dash for sanctuary.

We all emerged at the other end of the course filthy and exhausted as only an adrenaline-fueled piece of exercise can make you. The M-60 machine guns continued to crack, sending glowing dots of red down range. The thought was that between each of those bright red dots were four bullets without the red phosphor embedded in them. The controlled bursts didn't stop until the last man had cleared the course. We marched back to the company happy to have cleared this hurdle. As we marched, one of the drill sergeants led us in a particularly bawdy marching chant, and we picked it up with enthusiasm. We always marched to different chants whenever we were moving in formation, but this was a new one that had been saved for this evening. It was like an acknowledgement that we had passed a milestone toward manhood.

The completion of basic was signaled by the three measured events. First was the target range where we would be shooting at silhouettes of various distances. The goal was to measure the proficiency of each man with his main weapon. A rifleman that can't hit the targets isn't worth much. I finished the course with a score that gave me the sharp shooter's badge. I had hoped for expert but fell several marks below that ranking.

The second test was the obstacle course. Various obstacles had to be cleared in a required time. You had to cross a long pole over a mud pit on you feet. Then you had to low crawl under a web of rope. Then you had to crawl through a culvert. You had to swing across another mud pit. And the final obstacle was a vertical wall. Most of it I had no problem with, but the vertical wall had been my nemesis from the start, and the day of the test was no exception. I made it over and finished in the middle of the pack thanks to two of the drill sergeants yelling at my heels and threatening physical abuse as yet undiscovered if I failed them.

The final test was the one-mile run. We would all run and be timed. Any who failed to cover the four laps of the track within the minimum would be bumped back and have to retake basic. At least that was the threat. The Cornell track star kept talking about how the post record would be in jeopardy and he would have refreshments for us all when we finished. We started the run at a slow jog like we usually did in our daily runs. Then slowly at the direction of Sgt. Kia, the pace picked up. The track star broke from the front at an even stiffer pace. When he did that, something in me kicked in. I started to run ahead of the pack too. The track star was already a good eight seconds in front of me when I passed the front of the group. I kept up with his pace, but I couldn't seem to shorten the gap.

The track star was pushing even harder when we ended the second lap, and Sgt. Kia was yelling at the two of us to slow down or we would burn ourselves out. This only pushed the track star harder. When he pushed harder, I pushed back. I kept the same gap. When we turned the last curve to enter the last lap, suddenly Sgt. Kia was yelling for us to keep going; he thought we might set the post record. The track star pushed out even harder, and I found it within me to kick it up and keep pace with him. As we entered the final turn, we were starting catch up with the stragglers, and Sgt. Kia was exhorting us to push a little harder.

We didn't set the post record. I don't know how close we came, but we did run a good course, and I proved the point to myself that I had it in me to keep up.

Chapter Three
A WOC is Something You Throw at a Wabbit

At mid-morning the day after basic training graduation, those of us going to flight training boarded two large motor coaches. There was excitement in the air as we relished the completion of basic training and the expectation of learning to fly. In late afternoon, we pulled into Dallas. We were given forty-five minutes to eat and stretch our legs. Our appetites having been satisfied, we loaded back up and headed west on US 180 to Mineral Wells and Ft. Wolters.

We arrived after dark, and a chill November wind blew raising dust and sending a lone tumbleweed flying through the main gate with our buses. We circled the parade ground, and all eyes focused on a cluster of men standing together under a bright area light. I heard someone murmur, "Holy shit," as the bus pulled to a stop and the doors swung open. We all piled out and began the effort to retrieve our duffel bags. A loud voice commanded us to move with a purpose.

We were ordered to form up. Once we were lined up and dressed right as we had learned through endless drill in basic training, the training officers set about lining us up according to our height. Those of us who were taller were segregated from all of us of average high. After getting us separated, we counted off. There were six of us at the end of the line who were again separated from the others. We were taken to a barracks and directed to select bunks on the second floor. Everyone else was herded into two other barracks. Training advisers (TAs) followed yelling orders and generally harassing the new WOCs.

We six got to the second floor of the isolated barracks and made up our beds. Through the night, we watched the other two barracks as the lights in them stayed on and occasional yelling echoed over the sound of the cold Texas wind.

"What the hell are they doing over there?" someone murmured.

"Don't know, but it can't be good," came the response from one of the others.

We all tried to get some sleep and wondered what our fates would be as the six extras. In the morning, we were roused by a T.A. We joined our weary comrades and peppered them with questions about what had gone on during the night. Where the mess hall in basic had been a cacophonous chamber, this mess hall was deadly silent. We were instructed to sit with our backs away from the back of the chair. Our meal would be a square meal. Food was to be elevated vertically to the level of the mouth and then horizontally to the mouth. The utensil was to be then returned to the plate by the same path. The fork and spoon were to be seated at the left and right of the plate inverted. The knife was to rest at the top of the plate with the blade pointing to the left and the blunt edge out away from the dinner. In this manner, we ate our breakfast, or as much of it as we could with the "square meal" strictures and the limited time allotted.

Once we had finished our breakfast, we were taken to supply. There we all got our collar pins and shoulder patches. We were all instructed in proper etiquette in all circumstances. All responses to the TAs were to begin and end with sir. Proper location of the collar pins was given. We completed paper work, and each of us was promoted to Specialist 5. We would be getting additional pay, which everyone greeted with surprise.

Several men withdrew after the first week under the constant pressure applied by the TAs. One guy went AWOL. I would not have suspected the guy who went AWOL. He was on the small side, but he was wiry, and his dad was career army. I'm not sure why he went over because he had talked endlessly about flying and out ranking his old man. It took three days before the six of us were folded into the regular barracks. Even though we were now privy to the regular harassment, it was good to be included with everyone else. We began classes on aviation. The basics of the aircraft, the TH-55, were reviewed. We learned the use of the flight calculator. Basic navigation and proper conduct at the heliport and the staging fields was taught. In between classes, we had physical training and doses of petulant harassment.

We were made to learn inane things like the correct response to: What is a warrant officer candidate? The correct response was: A warrant officer candidate is a select applicant member of the armed forces undergoing intense military training to become an officer and an aviator in the United States Army. There was another question that would be posed to the candidates: What is a WOC? The correct response to this question was: A WOC is something you throw at a wabbit. This seems simple enough to master, but when you are surrounded by three TAs all yelling different questions at you in staccato, the simple becomes very complex. What's your service number? What is the diameter of the main rotor on a TH-55? Do you like Mr. Johnson better than me? What's the glide ratio of a TH-55 when auto rotating? What's the correct heading when flying straight south with a 20 knot cross wind? When you didn't respond correctly or you ignored one of the TA's questions to answer another one's question, you were down on the ground doing pushups. All the time the TAs yelling, "Do you want to quit? You do pushups like a little girl. Give me another twenty!"

Sleep deprivation was standard for us at this point. Fatigue compounded by the constant harangues of the TAs had an extraordinarily wearing effect on a person. You never knew when you would be singled out for a mobbing session. Somehow we had completed the first four weeks of our training as Christmas approached. The word was out that once we began flight training, the flight surgeon would require that we all be allowed eight full hours of sleep. That made everything that came more tolerable. We were all sent home for the holidays with the realization that when we returned, we were going to fulfill our ambitions.

Our barracks had the tail boom from a BellH-13 Sioux under it. The thing had gone unnoticed until the night before leave. The other barracks' TA decided that he wanted them to create a Christmas tree. Word came quickly that the other barracks wanted our tail boom. From nowhere a chain and a pad lock appeared. We lashed the boom to one of the cement footings of the barracks. When the TA discovered our action, he had us fall out. He called us to attention and demanded that we turn over the key. To the man we all denied knowledge of the keys' location. Clearly one of us was lying—a breach of conduct.

When we would not give in to the glowers and the threats, we were marched to the parade grounds and ordered to run around the track that circled the parade grounds. As we ran we began a chant, "We got the tail boom, boom, boom. We got the tail boom, boom, boom." On the far side of the field, the field lights were not on so we were running in the dark. One of the guys fell out and ran to a bank of pay phones to call our TA.

We had run four or five laps before our TA showed up. He talked to the other TA, and then he had us stop.

"Men, I appreciate what you did. If we were going to do something with the tail boom I would be all in with you, but the fact is we aren't going to do anything, and they have a plan. So we are going to give them the boom, and they are going to put it up between the two barracks so it will be both barracks' decoration."

There was no dissension in the ranks. When we were asked who had the key, we all responded that the key was in the barracks. The key was retrieved, and the boom was released to the other barracks. It was the close of our first phase of training, and we were all released for holiday leave.

Home was a bittersweet experience. One night at Ft. Polk, after a two-hour phone conversation, my girlfriend had accepted my proposal of marriage. A less romantic, more inept proposal has never been proffered. Why she said yes has astounded me to this day. One of my first projects was to get an engagement ring. We went to a local jeweler and purchased a modest ring. We spent Christmas at my grandparents' farm. I tried to ignore the fact that I would be heading back to Texas in a short period.

Going to the airport was another agony as I said goodbye to my newly extended family. Tears didn't help the cause at all. My flight was uneventful, and the bus ride was long. I was experiencing a whirl of emotions. The return to the jackals was not something to be eagerly anticipated, but the prospect of getting behind the stick of an actual helicopter was to be relished.

Upon our return, we were issued flight helmets, OD green nomex flight suits, silver-green flight jackets, and aviator sunglasses. Looking in the mirror, I looked like a wingless dragonfly, and I felt like the baddest man on the planet. We had been instructed on how to properly log each flight; this required multiple colored pencils. Most of us went to the PX and purchased a special pencil with four different colored leads that filled the bill and cemented our impression of our complete pilot look.

Every day was now centered around actually learning to fly. The first requisite was to find the hover button. Half of us would fly from the heliport to the various staging fields scattered out from the heliport. The other half would be bused out to the staging field to wait our turn at the stick and cyclic. The heliport was a large cement pad with rows of painted squares. Each TH-55 Osage, a large bubbled cockpit that resembled, more than anything else, a huge orange guppy, sat on its square. Across the southern border of the pad ran a set of painted squares. When leaving the heliport or when returning the helicopters, you would come to a hover over these squares and then would move to an open parking square where you would set down and shut the helicopter down after positioning it with the wind blowing at the left rear of the cockpit.

The staging fields were simple affairs of three concrete lanes. Each lane had four squares painted on it. The first three were for practicing approaches, and the first square was for taking off. I was in the half that was bused out to the staging field. We stood and watched as the TH-55s came into the rectangular flight pattern and made their individual approaches to the pads. Wind had a lot to do with flying, more than I had anticipated. Direction of the wind determined the traffic pattern. Helicopters would fly a down wind leg parallel to the three lanes. Then you turned at a ninety degree angle. Once you were at the end of one of the lanes, you turned and depending on how many aircraft were ahead of you, approached one of the three squares.

I watched, transfixed, as the individual helicopters bobbed in perfect hovers. Then one by one, the helicopters began to gyrate like flagging gyroscopes. Some would almost go vertical before suddenly snapping back to a perfect hover when the instructor pilot would take control. The instructor would relinquish control, and the helicopter would once again begin its mindless gyrations only to be snapped back to a perfect hover. This went on for half an hour. Then one by one, the helicopters sat down on the pad they had been hovering over and shut down. It was my turn. I was sure I would master this feat without difficulty.

Captain Guthrie introduced himself as my instructor. We went through the preflight together, and he emphasized the importance of each checkpoint.

Looking to make sure the Christ pin was in place and properly attached he said, "That breaks or comes loose you're screwed. There ain't nothing left to do but start an introduction to the man." Once we had checked all of the points, we climbed into the cockpit. He had me start the engine and then engage the clutch. As the rotor began to spin, the RPMs bled off from the engine. I disengaged the clutch, and the RPMs climbed back up. Again I engaged the clutch and the rotor picked up RPMs. Through several engage-disengage actions, I got the rotor into proper speed. Captain Guthrie then got the helicopter into a perfect hover. "When I say I've got it, you need to let go. No hesitation. You understand?" came his voice over the headset.

"Yes, sir," I responded.

He had me place my right hand on the joystick and then my left hand on the cyclic. At its end the cyclic had the throttle that controlled the engine RPM. When pulled up the cyclic controlled the pitch of the rotor. This was a delicate balancing act. As you increase pitch by raising the cyclic, you drain off power so you have to increase the throttle to compensate for the power bled from a greater pitch on the rotor blades. Last he had me place my feet on the pedals. These controlled the pitch of the tail rotor. The left or right pedal turned you around the compass. He had me move the pedals and see how the helicopter responded.

"You ready?" the voice on the intercom asked. I nodded my assent, and the voice said, "Son, you have to verbally respond to everything. I ain't going to be watching your head bob."

"Yes, sir. I'm ready." It was good that my helmet covered so much of my face so he wouldn't see my red face. Once I had the controls, I had us nearly vertical in the snap of an eye.

"I got it!" came the commanding voice over the intercom. I let go and trusted in Captain Guthrie's ability to right the aircraft. As quickly as I had gotten us into a perilous situation, he had us righted and back to a perfect hover. "Relax. Don't fight the controls. That hover button is right there. Just let it come to you. Let's try it again. Get your hands on the controls. Got it?"

"Yes, sir." I had my hands properly placed and was ready to beat this bitch into submission.

Again I had us at a perilous angle to the ground in a heartbeat, and again Captain Guthrie seized control and righted the helicopter. This happened a dozen times before Captain Guthrie announced it was time to head back to the barn. My clothes were soaked through with sweat, and I was more than ready to call it a day. I watched as he effortlessly flew us back to the heliport.

For three days I went out to the staging fields and watched as one by one my fellow candidates mastered the art of the hover. I had just finished placing the helicopter in a particularly awkward angle, and Captain Guthrie again righted it.

"Son, you're fighting this fight too hard. You need to relax and let the aircraft do its job."

There was a long silence then the voice came back on. "Pretty day, ain't it?"

I had not paid any attention to the quality of the day at all. I was focused on was how close I was to being washed out having never mastered the most fundamental part of helicopter flight. I looked around outside the cockpit and realized he was right; it was a particularly beautiful day in east Texas for January. "Yes, sir."

"Alrighty, then you have to controls?"

"Yes, sir," I acknowledged. I had the controls, and the helicopter sat in a perfect hover. This wasn't possible, just like that? I looked over to see if Captain Guthrie's hands were flying the aircraft. His hands were folded across his chest, completely relaxed.

"And that is where the hover button is," came the reassuring voice over the intercom. "Damn things hard to find sometimes, but once you locate it, you won't lose it. Alright, I need you to set down and shut her off. I got another man that needs help to find that thing." I lowered the pitch and allowed the helicopter to settle to the ground. I was amazed. One moment I was no closer to successfully hovering than I was to solving a major mathematical theorem, and then I was hovering like I had been flying all my life.

Once I had discovered the hover button, flying became everything I had hoped it would be. But my skillset fell short of what I had felt sure it would be. I had been among the last to master the art of the hover. I was also among the last to solo. I watched with envy as each afternoon our bus would stop at the Mineral Wells Holiday Inn pool as those who had successfully soloed were physically removed from the bus and thrown into the pool. As the number of us who had not soloed grew smaller, my apprehension swelled. What had begun as a pure joy was now a torture as I stressed over my failure to be released to fly on my own.

Finally I managed what had become the seeming impossible dream. On February 26th, 1969, Captain Guthrie, my flight instructor, gave me clearance to solo at Vung Tau staging field. Chlorinated pool water would never be as wonderful as it was that afternoon. I didn't even notice how cold the air was. After soloing there were occasions where we were allowed to take off by ourselves from the heliport and fly solo to the staging field. On a stormy afternoon, I took off. When you flew from the heliport, you would hover over to one of the pads. You would turn to the tower, and if the green light was lit, you were clear to take off.

I followed procedure and saw the green light. I took off gaining altitude as I headed to the staging field to the east of the heliport. As I turned my course to the east, rain cut loose, and the wind picked up. For some reason, the weather had no effect on me. I was just happy to be flying on my own, and I was more interested in maintaining a correct heading. The last thing I wanted was to over shoot my goal and land in a park in Dallas. When I got out to the staging field, I saw that there were only a handful of helicopters, and they were all sitting on pads and no one was sitting in any of them. Once I had made my approach, I was ordered to land and shut down.

It turned out that I was not supposed to have taken off at all. The light had been turned to red immediately after I took off. At first I was lectured for having ignored the red light. Later that day, it was confirmed that the light had indeed been green when I took off. There had been a delay in turning it to red. My instructor asked me if I hadn't been a little nervous flying in a storm. I told him that honestly I was too fixed on getting to the staging field. He laughed.

"Navigation will keep your attention. Sometimes that's a real good thing."

Every day we went to the staging fields to practice the various flying techniques. We practiced autorotation, the technique used for an aircraft that has lost power. We learned techniques for when the tail rotor failed while you were in a hover. We practiced maximum performance takeoffs and landings along with normal landings and takeoffs. I had a little over forty hours in the TH-55 when it was time to make my check ride. A different instructor pilot would take a ride with me to certify that I was ready to advance to the next phase of training.

The day of the check ride, I was so nervous I could hardly breathe. All of the other guys who had passed the check ride assured me I would do fine. I didn't. That afternoon I was told that I had not passed the check ride. I would be recycled to get more time in the helicopter before I took my next check ride. The recycles were a small group, and the structure was nowhere near as strident as it had been in the regular flight. The issue now was too get us proficient enough to move on in our flight training. My new roommate was a guy name Malcom. He went by Ted. I found him to be extraordinarily congenial.

Ted had been in the army for a while and had served in Germany. He had even been married to a German woman. The marriage had not lasted because as he said, she was extremely hardheaded and unwilling to concede on any issue. Ted and I were kindred spirits, and we were both determined to succeed in our second chance. We spent a lot of our time attempting to bolster each other in our efforts to become great pilots.

When all of the helicopters returned to the heliport after a training flight, they stacked up in long descending lines toward the landing pads at the heliport. Like orange pearls on an invisible sting, the TH-55s would approach the heliport, everyone trying to keep proper spacing and making a smooth, even descent to the heliport. I was flying solo back to the heliport, trying to monitor my instruments and keep an eye on the helicopter ahead of me, when all of a sudden the line I was in fell into complete chaos. Orange bubbles were breaking in every direction, and the radio came alive with chatter. Voices were demanding to know if anyone had gotten the son-of-a-bitch's tail number.

I had not seen it, but apparently a navy jet trainer had roared out of the west just above the deck right through the returning flight. The helicopters were just beginning to reform when the trainer returned, this time out of the east. Again the helicopters made their manic attempt to avoid the oncoming jet and each other. The radio again came alive with instructors demanding the tail number. By some miracle, no one was injured, and no collisions occurred. To my knowledge, the culprit was never identified, and the ire of many instructors went unquenched. While none of the instructors found humor in it, several of us students thought it funny, especially since no one had been hurt and we had all gained some instruction in avoidance flying.

Along with the WOC candidates, there were a large number of new second lieutenants who had just graduated from OCS (Officer's Candidate School). Once they made it through OCS, they were assigned to a specific unit stationed in Viet Nam. Since they were assigned to a unit but were still in the United States, they were being paid separate rations along with flight pay and base pay. This made their paychecks ridiculously large. One day I was taking off from the heliport when I looked down. The parking lot in front of the heliport was packed with Corvettes. It looked like the parking lot outside the GM assembly plant. I had never appreciated their windfall until that moment.

We had been in the recycled flight for three weeks when one night after dinner, Ted told me he thought perhaps not all of us were meant to complete flight training. My mind would not accept that thought. I assured him that being recycled was merely a momentary setback. We would be graduating and flying in no time. Ted was skeptical. He clung to the idea that he was just not meant to complete flight school. At last he wearied of the argument. "I'll give it a few more days" was all the commitment I could extract from him

Two days later it happened. I was flying solo. I was attempting to master the max-performance landing, an action that requires the helicopter to approach the landing pad at a steep path as if it's coming in over tall obstructions. You didn't actually land; you just came to a hover over the pad. I overshot the approach pad and came to a hover half way between the second and first ones. I looked over to see a TH-55 laying on its right side. The three rotor blades were horribly twisted, and a knot of men in OD flight suits were collected around the mangled mess that had been a fully functional Osage minutes before. Everyone flying was ordered to move out to the farthest strip and shut down.

I sat and watched as a big Sikorsky H-34 medivac helicopter flew in, and a couple of medics appeared from the side cargo door. They began working on the limp body. Once the man was extracted and on a cot, they maneuvered the cot into the bay of the medivac copter, and off it went. I found myself wondering who it was and how they would be. Once the medivac copter was away, we were all ordered back to the heliport. I flew back to the heliport with my mind clouded by thoughts of who it was and how badly were they hurt. The biggest question that came to me was: Would they be able to fly again?

Once I had parked my Osage, I walked to where the bus was picking us up. On the bus, one of the instructor pilots told us it had been Ted who had crashed. They were taking him to Laughlin Air Force Base Hospital. It was unclear how severe the injuries were; Ted was unconscious when they got to him. I played back in my mind what Ted had said two days before: "Maybe some of us weren't meant to complete flight school." I also thought about my arguments made against his quitting. He was hurt because of my mouth.

Ted never recovered from the injuries; he had basically been dead from the moment his helicopter flipped and crashed. The seat belt meant to hold him secure in the seat had apparently not held him in place. He had fallen forward enough that his head was trapped between the cockpit frame and the ground. The fuel tank mounted behind his seat had not broken away from the cockpit but had instead remained in place, spilling the aviation gas all over Ted as he lay trapped. The accident had been caused by Ted, for reasons unknown, burying the tail rotor into the ground. The result was an uncontrolled spin that placed the helicopter on its side.

The room we had shared seemed particularly empty now. I played the accident over and over in my head. What the hell had happened to cause Ted to pull back on the stick like that? I slowly came to the suspicion that I might have been one of the causes. I lost all focus, and I couldn't help this gnawing suspicion that I had had something to do with the crash. I could not say with certainty that I had, but I couldn't escape the nagging feeling that my overshooting of the approach pad had been involved.

In April I took my last check ride. One of the requirements for promotion was to perform a max-performance takeoff and landing. Instead of the normal forty-five degree approach landing and takeoff, it was more of a seventy degree angle. I had been having problems with these at the practice field, so I had been paying particular attention to these two connected techniques. I had thought I was gaining control over these procedures. Auto rotation had been my current bug-abo. But today, of all days, I performed a perfect auto rotation, barely sliding a foot after touching the ground. It was my max-performance takeoff and landing that were really awful. When we returned to the heliport and I had parked the helicopter, the instructor told me he couldn't pass me.

"Your max-performance landings and takeoffs were really iffy, and they are very important in the next phase." He looked over his notes. "Your normal approaches were good and your auto rotation just fine. Have you had problems with your max-performance landings and takeoffs?"

"Yes, sir, off and on."

"Well that's a vital component of flying, and I can't move you on without having that mastered. Unfortunately, they aren't going to give you a third chance. I am afraid this is it."

There was no lying my way out of it; they had probably been the cause of my roommate's accident. I felt drained. Everything I had believed myself to be, all I had aspired to, was erased like writing on the beach as the tide flows in. If there was a word for me it would have to be failure. What would my family think? What could I possibly do to overcome this failure? The answer was nothing. I was a loser and nothing could matter now.

I was transferred to a processing company to await reassignment. This company had been fairly unique for some time prior to my assignment. I was told that when men were transferred in it, it became a more or less permanent assignment. Men had gone AWOL for long periods without anyone noticing. Some guys were going to college and just showed up for paydays. No one was the wiser until someone stole an Osage and crashed it into the lake. In order to figure out who had done it, roll calls had to be taken. When a quarter of the men assigned to the holding company failed to respond to the roll call, the jig was up.

The result was a much stricter handle on those assigned, and the effort was actually made to get men reassigned to an actual duty concordant with the military mission of the army. I was ordered to an office where an older woman asked me if there was an MOS I would like to be assigned to. I remembered the scores I had had in language, and I asked about that. I was advised that would require reenlistment for another year. The army needed assurance that they would get full use of the training. I had no interest in extending my term in the army. I didn't know what would become of me, but military life outside of a cockpit was not among my choices.

Finally, I just asked to be assigned somewhere close to home. "We'll see what we can do," came her reply. I was dismissed to my barracks. Unlike most other duty stations, there seemed to be no ambition for making work projects at Ft. Wolters. We were engaged in policing our area and keeping our barracks clean. Twice a day there was roll call, an oversight corrected. Usually at the evening roll call, some of the washouts would be advised of a new assignment. One week after washing out, I was given my orders. I was to report to Ft. Knox for armor training MOS 11-E in army dictum. I found myself with an attitude of caring less. The only bright spot was that I had five days travel time. I could go home and tell everyone, "Yeah I washed out of my dream. The dud is home."

Chapter Five
11 Echo

Of course my family felt bad for me. There were no recriminations, no finger pointing. I spent two days at home and then booked a flight to Ft. Knox. I was now back in the real army. Formations, training, roll call, and military duties, like policing and KP. The training unit hotel 4-1 was a smaller one commanded by a gentleman, Captain Gray. He was of average height, had a fair complexion, and a stern demeanor. He was waiting orders to go over to Viet Nam. He spoke with a mild Southern drawl. Rumor had it that he was from a military family where most his ancestors and some of his siblings were career naval officers, and he had chosen West Point over Annapolis. One can only surmise what family reunions were like for him. The entire training unit was housed in one barracks. One thing that I had learned to be at Ft. Wolters was a strack troop. I kept my boots spit shined and my brass polished so I could blind a man at twenty yards on a bright day.

On that first day, as we settled in to the barracks, I was introduced to my new cadre. Asycue was already there when I rolled in. He was about my height and weight with a soft Southern drawl akin to our new CO. He introduced himself and gave me a quick resume of his military experience, which encompassed basic training at Ft. Polk and a flight to Ft. Knox.

Radler was the next to show up. He was shorter and a little stockier than Asycue and me. When Asyque made a comment about how he had gotten through basic with a little spare weight, Radler replied that anyone who loses weight, regardless of what the extraneous efforts of the army were, just didn't have his mind right. When liver was served for dinner in basic training, Radler, single-handedly, disposed of a majority of the liver prepared for his company. The mess sergeant was left with a rare conundrum; no left over liver for the ensuing dinners. Radler was committed to the proposition that nothing was inedible and throwing food out was a sin before God.

Valentine showed up toward the end of the day. He was a good two inches taller than anyone else in the training company; he was a slender, good-looking guy from Boston. His talent would make itself apparent while we were patronizing the Ratskeller, the base enlisted men's bar on Ft. Knox. Designed after the real brew hauses of West Germany, it served 3/2 beer to all soldiers over the age of eighteen. As much as Radler could eat, Valentine could drink. No can of 3/2 beer was downed without a comment, made loud enough and in his most erudite Boston accent, that it wasn't as strong as the water served in Boston taverns.

This loud disclaimer and the accent were certain to draw a response from some unsuspecting patron. When we were with Valentine, we rarely paid for any beer. He would wager that he could drink a can of 3/2 beer from the can in two seconds. Wagers quickly blossomed, and with two or three pitchers of beer on the line, Valentine would proceed to demonstrate his skillset. He would produce the P38 hooked on his dog tag chain and remove the top completely. Once the inevitable argument about whether or not the act of removing the lid had been part of the wager, Valentine would set about making the beer disappear. He could pour the beer down without swallowing even once. Time and again Valentine earned free pitchers.

The weird thing was that I never once saw him show any sign of inebriation regardless how many beers he downed. When he had had quite a few, he would wax poetic about Boston and how the only thing wrong with the town was that a part of the population was Irish. Those assholes walk around like they own the town and everyone else owed them for being allowed to live there.

For field training purposes, we were formed into crews of four men and assigned to an M-48. The M-48 was the current main battle tank of the US Army. It was a giant 45-ton tub of steel with a traversing turret. The primary armament was a rifled 90 millimeter main gun. This weapon fired a variety of shells for combating fixed positions or other armored vehicles. A coaxial 7.62 machine gun, and in front of the vehicle commander's hatch a M-2 .50 caliber machine gun, were for dealing with infantry or smaller vehicles. A huge continental diesel drove the whole thing on treads that ran the length of the body on each side.

We were taken out to a driving range, and each of us was given an opportunity to drive an M-48 around an obstacle course. There was the requisite mud pit that each of us had to negotiate. The expectation was that the steep bank and the deep muddy water at the bottom would come boiling over the front of the tank and into the driver's hatch, soaking the driver. I managed to avoid this calamity for two reasons. First I was one of the last to drive, and the others had taken on so much water the puddle at the bottom was much shallower, and two, I took the slope at an angle avoiding putting the whole front deck into the puddle.

A man from Mississippi, Kyle, was made our barracks officer because he had some ROTC experience. That seemed to be his only qualification. He was as close as anyone I have ever met to being a Coca-Cola addict. Every morning before breakfast and then at least five times during the day, he would go to the soda machine outside the company headquarters office, plug in his fifty cents, and get a coke. The next thing was something I had never heard of, let alone witnessed. He would buy a small bag of peanuts and then pour the nuts into the coke bottle. This brew he would then chug.

I was again one of the few soldiers that knew how to drive a stick. That ended me up on guard duty more than anyone else. I didn't have to walk a post, but I did have to drive the sergeant-of-the guard around the guard posts all night.

We had classroom where we learned maintenance, radio operation, more orienteering, and tactics. We spent time in the motor pool learning how to remove damaged sections of the track and how to short track if we ran over a mine. The treads were driven by a pair of cogged drive wheels mounted high, just behind the engine in the rear. These drive sprockets were what pulled the tread in a circular motion. If a section of the tread was damaged, it could be removed, and the vehicle could move on a shortened tread. This left the forward wheels idle. A pair of clamps fit into the holes in the track that allowed the drive sprocket to pull the tread. These clamps had threaded rods that contracted the clamps' length, pulling the slack from the tread. Once the tread was pulled taut enough, slack was created between the two ends of the clamp to allow the pin that went through each section to be extracted. Once a section was removed, any maintenance necessary to the tread could be made.

The forty-five tons of metal had a torsion bar suspension that we had to learn to remove and replace. Each of the four crewmen took their turn extracting and replacing a torsion bar. As the man up struggled to extract and then replace the bar, the other trainees took their turn making every kind of obscene suggestion regarding the one doing the maintenance's success. The sergeants who trained us were not nearly as tightly wound as what I had experienced in basic or at Ft. Wolters. One beefy sergeant whose girth barely allowed him to fit through the commander's hatch would wax poetic about the army and how a career in the military wasn't all that bad. He hoped to make it to twenty years, but first he had to win the battle of the waistline. He would be mustered out the next year if he didn't meet fitness standards.

We had been in training for three weeks. I had just come in from my third course of guard duty and hadn't even gotten a chance at a shower when I was called down to the company office. I reported to the first sergeant. He directed me into the CO's office where another soldier that I had not seen before stood at attention. I looked at the first sergeant to get confirmation that it was alright to interrupt. He nodded and jutted his chin toward the door. I knocked at the door jamb and was ordered in by Captain Gray.

"Come in, Private," he ordered in the soft drawl that characterized his speech.

"Sir, Private Swartz reports as ordered."

He looked around me to the open office. "Sergeant, get Private Swartz a side arm and a clip."

The first sergeant stepped into the doorway that I was partially blocking. "A clip sir?"

"That's right, a clip. And make sure it's full." Captain Gray's tone remained smooth and even.

The first sergeant pulled me out and went to the safe. He extracted a .45, a holster, and web belt. He handed it to me and instructed me to put it on. I got the web belt hooked up and then felt the holster to confirm that the pistol was there. The sergeant then took out a clip and, noting that it was full, handed it to me. Properly equipped, I returned to the captain's office.

Captain Gray could see I was holding the clip in my left hand. "Put the clip in the pistol, son. It isn't going to do you any good in your hand."

I lifted the leather flap and extracted the .45. Making sure the safety was on, I slid the clip up into the handle and gave it a pop with the butt of my hand. Once I was satisfied that the clip was in place, I started to replace the weapon into the holster.

"Hold on! Chamber a round." The voice was now more authoritarian in its tone.

I looked around to see if anyone else was hearing these instructions. There was just Captain Gray and the private who still stood at attention. For the first time, I noticed that his fatigues were soiled and his pants were not bloused. He seemed indifferent to the activities swirling around him. I looked at Captain Gray.

"Go on. Chamber a round, and make sure this young man over here hears it going in."

I lifted the pistol to my right ear and drew the slide back. I released the slide, and the spring pushed the slide forward, pulling the first bullet out of the clip and into the chamber. I lowered the pistol slowly as if the gun could discharge of its own will. As the pistol passed in front of my eye, I again noted that the safety was on. I took the safety off and squeezed the trigger as my left hand held the hammer in place. I allowed the hammer to go forward and then put the safety back on.

"Now, Private Swartz, I want you to escort the private here down to battalion where he is going to be court-martialed for being a chronic AWOL. If he attempts to make a break for it, and I hope he does, I want you to empty your clip into the back of his head. Do not tackle him or fight him into submission; just put him out of his misery! Do you understand?"

It took me a second to realize I was being questioned. The idea of shooting a fellow soldier regardless of circumstances was a whole new horror. Once I realized the silence enveloping the office required my response. "Yes, sir! I understand."

"Good!" He smiled at the AWOL standing before him at a poor imitation of attention. "I will not have someone like this mutt placing a black mark on my record and then be walking around smiling about it. Since I have made myself understood, you are dismissed."

Nothing could have been less clear, but I answered in the affirmative. Having my orders, I waited for the private to leave, and then I followed in close proximity. Once we were out of the building and walking past the mess hall, I told the kid I was escorting not to do anything stupid. His response was that he had no intention of causing trouble except in the court-martial where he was going to provide the hearing officer with an ear full.

At battalion the hearing went quickly with a decision that the AWOL would be condemned to ninety days hard labor. He was returned to my custody so that I could escort him to the Ft. Knox stockade. A duce and a half and driver were provided to get us to the stockade. After I had signed him over to the MPs and gotten back into the back of the truck, I found myself a confirmed conformist. I would follow the rules with a smile rather than face the consequences of misbehavior. Relieved of my responsibility, I suddenly realized that I had a loaded pistol at my side. I removed the pistol from the holster. I released the clip from the handle and then removed the round from the chamber. I put the round back in clip and then the empty pistol back in the holster.

Clary was the next to encounter an event outside the normal areas of expectation. He was walking fire guard when he found two of the trainees in the latrine. A kid from Wisconsin who was always saying "Baby Moons" in a child's voice was on the receiving side of the amorous probing of a rather large kid from the south side of Chicago.

At first Clary was reluctant to go into any details. But, with constant inquisition by all of his fellow trainees, he finally came forth with the details. When they were caught, the Chicago kid begged Clary to just let it go. Clary, after a short period of cogitation, thought better of it and advised the first sergeant. It took almost no time for the two lovers to be packed up and shipped out. After the third telling of Clary's encounter, details about the event began to become somewhat enriched. The foaming of the soap they had used as a lubricant and the grunting of the two as they proceeded in their tryst grew more detailed. All causing raucous laughter from all those hearing the story.

The one thing that always made me suspicious of the whole event was a comment I overheard from "Baby Moons" as he was packing up his belongings. It was to the effect of "Yeah but I'm going home and look where you guys are going to be going."

The guy he was talking to sputtered, "Germany?" The suspicion we all had was that we would be assigned to Germany 'cause there was no need for armor in the land of green.

Baby Moons response was, "No, Viet Nam. That's where they're sending you guys. That's why they're going to give you all additional training after. I overheard them talking about it." He would say no more about it, not who had said it or when he'd heard it. For the most part, it was disregarded by all of us as someone making excuses for poor behavior.

Louisville called to Asycue like the Sirens called to Odysseus. He was powerless against its call. The difference was the rest of us, while we didn't hear the call, had no mast to lash Asycue to. Once the weekend came and his duties were done, Asycue was cleaned and ready to catch the shuttle. Radler and Clary were almost always with him. I went several times, and it was always interesting. The trip was not a long one, but Radler was incapable of making the trip without sustenance. We would stop at a pizzeria just past the mid-point, Radler's point of no return. The place was popular, and there were almost always half the tables filled even when we got there late.

We had managed to clear away two large pizzas, Radler having eaten one almost completely by himself. Asycue announced that it was time to get going. As we filled out, we passed an empty table that hadn't been bused yet. Four pieces of pizza cut into squares sat on an aluminum serving tray in the table's center. With a deft spin, his right arm shielding his left arm's action, Radler scooped up the four pieces. Two disappeared before we were out the door.

Once outside Asycue turned to Radler. "What the hell? You're eating someone else's pizza?"

Radler looked the innocent except for the bulging right cheek. "Whut?" he offered through a wad of pizza crust, cheese, and an unknown combination of toppings. "They left 'em on the pan, and those pieces hadn't been touched. Why waste them?"

"You practically ate a pizza by yourself, and then you hoover someone else's pizza. I mean sweet Jesus, Radler. I don't want you anywhere around if I get killed. You'll be gnawing on my ham before I go cold."

A big grin seized Radler's face. If he was on a mission to gross out everyone, he might just as well go all in. "Few things better than a cold ham and cheese with a little mayo." Getting the groans he'd expected he added, "Course the cheese has to be Swiss, that goes without saying."

"Sick son-of-a-bitch!" was Asycue's only response. He had been bested, but there were adventures to be had in Louisville. We had made our way down the drag about a half a mile from where the shuttle let us off. The usual discussion was being had about what everyone wanted to do. Asycue was uninvolved. His entertainment compass was yet to settle on a true course. Finally, Radler announced that he was officially sick.

"Really? How is that possible? I mean, just because you ate half of Ft. Knox before we left, I can't imagine that would have an ill effect on you, Kenny."

"No shit, I'm serious. I'm feeling a little cramped in the gut. I may puke right here," Radler groaned.

Not wishing to be just another group of Neanderthals from Ft. Knox, we began looking for a place for Radler to get relief in semi-privacy. Someone spotted a hotel up the street. Hustling along with two of us on either side of Radler supporting him, we moved en mass. We entered the lobby and began scouting for the bathroom. Clary spotted the bathroom just past the two elevator doors and directed Radler with a tug on his right arm. Two of us idled in the lobby while Radler and his supporters made the rush for the bathroom, slightly hunched over with abdominal discomfort.

He had disappeared less than a minute when a roar erupted form the hallway. "Jeez, I'll bet that one registered 4 on the Richter," I opined.

"Well it wasn't Krakatoa, but it was pretty embarrassing just the same," Asycue replied.

We all tried to look nonchalant as we waited for Radler to emerge from the bathroom. At last as our loitering was reaching the uncomfortable stage, Radler came out wiping his chin with the remains of a paper towel.

"You ok?" Asycue asked.

"Yeah, but I'm feeling kind of empty." Radler looked sheepish. He had managed to vacate his stomach without getting anything on his clothes, but now having emptied himself in the Holiday Inn bathroom he was hungry. We all sniggered as we left the lobby and emptied out into the street. We spent the balance of the weekend idling, with nothing particular to do and all the time in the world to get the job done.

There was a pair of characters who lacked familiarity with the real world. One was a kid, Daily, who was variously sixteen through nineteen depending on the day and his mood. He was short and stout and, as I said, unfamiliar with truth or reality. The end all for him came on a Sunday night. He came into the barracks more disheveled than usual with a pair of handcuffs dangling from his right wrist. He began mumbling about a run in he had with the local constabulary. He claimed he'd gotten nasty with one of the cops, and when he tried to arrest the kid, he had kicked him in the crotch and run off. The problem was that the cop had managed to close one of the cuffs on his right wrist before Daily had made his escape. He held up his right wrist to illustrate his story. I looked, and it was clear that the handcuffs were a toy pair you would buy in a five and dime in one of those sheriff's deputy toy packs.

Being the only one in the barracks, I was forced to address the issue. I tried to point out the holes in his story to no avail. Daily insisted on going into the company office and reporting his misdeed. The sergeant on duty immediately saw through the story and told him to go back to the barracks where he was sure someone could remove the toy handcuffs. Daily refused the direction, and the end came with a phone call to the CO.

Our captain was about to be shipped out to Viet Nam. He was close to being shut down as a state side officer and going into combat. He saw our company as a satanic attempt to sabotage his career. He showed up in the office and summoned the kid. The handcuffs still dangling from his wrist, Daily went to the office. He didn't return that night. Several days later, Daily announced that after training, he was being assigned to a unit in Germany. Quite the punishment and no one was certain of what had happened, but Daily was the quietest of individuals and the fabrications stopped.

The other prevaricator was a guy named Cox. His specialty was one-upmanship. If anyone had done anything special or unique, Cox could be counted on to come up with a story about one of his experiences that eclipsed it. He also considered his knowledge to be encyclopedic. I found just his presence to be annoying so I avoided him with studied intensity.

As we approached the end of our training, there seemed to be a growing sense that the next assignment for us was in Germany watching the Russians watching us or in Ft. Reilly playing at tank warfare in the flatlands of Kansas. Either prospect didn't seem too bad given the alternative of Viet Nam. The thing was, as the majority opinioned, there was little need for armor crewman in the jungle. I fell to this line of thinking, and so I found my attitude focused by this logic. There really wasn't any need to get into too big a twist about the balance of my tour of duty. I was fast approaching the first anniversary of my enlistment, which meant I had but one year remaining.

One of the last events in our training was bivouac and gunnery. We went out to the range with the tanks we had been training on. When word went out that we were going to the field for three days, Kyle set about assuring himself of a study supply of Coke. Storage was a problem he overcame by using the main gun of his assigned tank. We got out to the range, and the first duty was to unload a flat bed of main gun ammunition. In the summer Kentucky sun, we sweated our backsides off taking large wooden crates of 90 millimeter ammunition off the truck and stacking it back off the firing line.

We had finished, and Kyle had been the first to return to the bivouac area and sprawl out on the cool soil under a cluster of shade trees. He looked at total peace until someone told him his tank was the first pulling on the line to test fire the main gun. "NO!" erupted from his mouth, and off he ran toward the range. Too late he got to the range. The vision of cans of Coke sailing down range filled his view. Not only did he lose his supply of Coke, but he also shed a portion of his posterior when the first sergeant found out what he'd done. Pointing out he had the Coke cans lodged in the barrel instead of flying down the range a major catastrophe could have ensued.

As it was, by some miracle, major damage to the barrel was averted. Kyle for his part spent several hours cleaning up the main gun and prepping it for use the next day. The fiasco did nothing to diminish Kyle's thirst for Coke. The machine outside the office was his first stop when we returned from the range in two days.

Our training was finished, and we went nowhere. Orders came, and as if the army was interested in proving how pointless speculation was, we were assigned to the very same unit we were in. We were to undergo additional training on a special light tank: The Armed Reconnaissance Airborne Assault Vehicle M-551 Sheridan.

Chapter Six
2 TANGO

Ascue was certain that our assignment to additional training was meant solely to discourage us from speculation and the circulating of rumors. When we found out that the Sheridan was armed with the Shillelagh missile, an anti-tank weapon designed to hit anything you could see, our belief that we were destined for Germany seemed certain. There was no way they would deploy such a vehicle in the jungles of Viet Nam where you couldn't see fifteen feet in front of you.

The Sheridan was a wicked fast tank. The six-cylinder Detroit Diesel engine could move the seventeen tons of aluminum at forty-three miles an hour. Its armament included a 7.62 MM coaxial machine gun, M-2 in front of the track commander's cupola, and a honking big 152MM main gun. This mammoth weapon fired conventional shells that had special paper casings for the accelerant. The shells it fired were the standard HE (high explosive) HEAT (high explosive anti tank) and canister (an anti-personnel shell encasing 10,000 flechettes). Once the round was fired, there was a compressed air canister that would blow into the chamber to push out any burning casing that had not been pulled out of the muzzle. But the main reason for the huge barrel was the Shillelagh missile. A see it shoot it, anti-armor weapon meant to give the Sheridan standoff capability against large, more heavily armored, main battle tanks. In addition to its speed and armament, there was its ability to be airdropped.

We watched films of the Sheridan being dropped from the rear hatch of a C-141. Three chutes deployed, and special packing covered the bottom of the tank. When the package hit the ground, it appeared that the cradle exploded and the three parachutes went limp. A four-man crew rushed into the scene and began to collect the chutes and undo the harnesses. Once the parachutes and harnesses were clear, the crew clambered aboard. Shortly a blast of exhaust flew up from the rear deck, and the tank moved off toward cover and away from the camera.

The next clip showed a C-130 flying low over an open field. The rear ramp was down. A parachute blossomed from the gap under the plane's tail, and as it filled, a Sheridan attached to the parachute rolled out the ramp, bounced several times, and spun to a stop. Again a crew of four appeared from off screen and set to freeing the tank from its harness. Once the packing was cleared, a gush of exhaust erupted from the back deck, and the tank lurched away from its landing spot and off camera. Everyone in the classroom was duly impressed. When the film was concluded, lights went on, and the instructor asked for questions.

A voice, not waiting for acknowledgement, asked the question on everyone's mind: "We gonna be doing that during our training?"

"No. That film was done at Ft. Benning." The instructor set the film to rewind on the projector as he surveyed the sea of faces. "As you can imagine, the airborne are much more interested in the air assault capability of the vehicle." He sighed. "I am not interested in jumping out of airplanes. If I was wanting to jump out of airplanes, eleven echo would not have been my MOS choice." A chorus of laughter erupted as everyone seemed to agree with the sergeant's point of view. "The purpose of this class is just to make it clear to you that the M-551 can be airdropped in multiple ways. We will be showing you how to rig the tank for those kinds of drops, but do not ever expect to be called on to make an air assault."

The last film showed a crew putting up the skirt attached to the hull. Once the skirt was up, the tank was driven into a river where it proceeded to cross using the treads to propel it. The agonizingly slow crossing ended with the Sheridan climbing the opposite bank and disappearing into the woods on the opposite bank. The film finished rewinding, and the end of the film flapped against the projector. "That concludes today's class. You may return to barracks."

All of the classes now concerned themselves with the particulars of the M-551 Armed Reconnaissance Airborne Assault Vehicle. We had already been trained in the basics of being an armor crewman. It was during a class on the care of the Detroit Diesel engine that I got into trouble. We were learning about the water supply and how to monitor the cooling system when a pair of C-123 Providers flew overhead. They had spray nozzles under their wings, and they appeared to be spraying a liquid. As we watched, the training sergeant told us they were doing mosquito abatement and the spray wouldn't harm us. One of the other trainees, Cox, said something about the two planes being C-47s. I corrected him with a reminder that the C-47 was a low wing monoplane and what we had just seen were a pair of Providers.

Cox was a know-it-all from way back, and when he took on that condescending tone of his and tried to tell me that I didn't know what I was talking about, I got my back up. The exchange between us started getting heated, and then the training sergeant stepped in. We were instructed to return to the company headquarters and tell the first sergeant of our misdeeds. Cox and I bickered all the way back to the company office. We walked into a room filled with chaos. Captain Gray had departed for Viet Nam, and the new CO had not reported.

I believe that Cox and I were the beneficiaries of the disorganization. The first sergeant told us that this incident would be put in our jackets and he would leave it to the new CO to provide suitable punishment. Since it was late in the afternoon and maintenance class would soon be over, we were sent back to our barracks. While the punishment was mild, I logged this in my memory as something that would require repayment, like a debt. One day Cox and I would have to have words.

The range was again a scene of armored assault as we rolled the Sheridans down to the range to learn the operation of the main gun. One of the cool things on the vehicle was that if the track commander kept the red pommel grip squeezed, the main gun would remain locked on the target down range. We had all hoped to fire a Shillelagh, but that was not to be. Our instructor told us that just the week before, a bunch of dignitaries had come down for a demonstration. A pyramid of 55-gallon drums had been erected down range. A couple of gallons of gas had been poured into each drum, and then they had allowed the summer Kentucky sun to ferment the brew.

On demonstration day, the wigs were seated in the bleachers anticipating a fireworks display to rival the Fourth of July. The Shillelagh took off down range, following its infrared guide. That is until it got about two thirds of the way to the target and it began to climb. Despite furious efforts to correct the trajectory, the missile kept climbing; it managed to clear the pyramid with a good ten feet to spare, and then it continued down range. The demonstration was over. They had been so sure of the Shillelagh's deadly accuracy that they had only brought one out to fire. All of the suits packed up and headed back to the base where conciliatory drinks were shared before planes headed back to Washington to vote on appropriations for the needed weapon system they had just observed fail.

The first thing that became evident was the power of the main gun. When fired, the recoil would lift the front deck a good eighteen inches. Quite a ride for the driver. When fired from the side, there was no resultant rock. The point being: Firing from the side was preferable. After each shot, the discharge of air and small pieces of burnt debris came out the muzzle. Each discharge was met with a roar of laughter.

We fired conventional rounds at the range, blowing up wrecked vehicles and improvised shelters. One of the things learned was that the fire ring that prevented blow back when firing a conventional round could get stuck to the breach block when it got dirty. If it stuck to the breach block, it would tear up the paper cartridge. The reason the ring was removable was when firing a Shillelagh, there wasn't enough clearance for the breach block to rotate over the end and close behind the missile—another failure of engineering.

Morale was pretty high in the company. The anticipation of deployment to Germany was palpable. That came to an end when we returned from the gunnery range. Orders awaited us all. We were to be assigned to Viet Nam. Asycue was certain this was another trick of the army. There was no convincing him that all decisions were made to fulfill the army's desire to keep the lowly GIs in the dark.

"You watch; they're messing with us. We'll get to Oakland, and they'll change our orders for Germany."

Radler sat on his bunk, a hang-dog look on his face. "I think you're wrong this time. I don't see them going through all this to mess with our minds. I think we're going."

The issue was argued back and forth until finally it was concluded that we would all find out soon enough.

We finished our training and went home for two-week leave. In the middle of August, we would report to Oakland Army Base for deployment.

Chapter Seven
LEAVE

I called my mom to let her know that I was coming home for a two-week leave and to arrange for a pick up at O'Hare the next day. The first question from her was where would I be going after leave. I told her, and there was silence for a long minute.

"We'll be there," finally broke the steely quiet.

"Thanks," was all I could muster. I was putting her in a dark place, and there wasn't even the compensation of being a pilot. I could have avoided all of this, but I had to be a pilot. There was no place of comfort, just the misery of a bad decision being lived through.

In 1969, O'Hare was a crossroads populated by lots of military personnel traveling to various assignments. I was just one more khaki-clad soldier moving through the maze of gates and counters. It had been decided that I should wait at my arrival gate until my mom, sisters, and Linda, my fiancé, showed up. When they arrived, there were hugs and kisses and then finally a mutual agreement that we should get going. We went to baggage claim where I retrieved my duffle bag, and then it was out to the car and home.

Mom guided the car out of the parking lot into the flow of commuter traffic, and there was an unusual quiet. Mom stayed focused on the traffic, and my sisters for a rare moment in time sat quietly in the back seat. The stillness didn't end until we arrived at the border of

Arlington Heights. At last my mom asked if anyone was hungry. All expressed the desire for something to eat. My mom offered the name of a local family owned restaurant, and we all accepted the choice.

The next day my mom dithered over whether to go to work or not. I assured her that I would be fine and she should go. The two weeks of leave compressed everything. It created a finite perception of time that had not been there before. Before being assigned to Viet Nam, there was all of the time in the world to do, to say, to be. Now, with each night's news being filled with the latest casualty figures and the grim pictures of wounded and killed, time was no longer infinite. Mom reluctantly went to work while my sisters, Linda, and I hung out at home. I had no agenda, just to have time with family.

When the weekend came, the five of us went to my grandparents' farm to see all my relatives. The farm had been the center of my family for as far back as I had memory. A large two-story, six-bedroom house— it had been built by my Great-Granddad Rush near the center of 800 acres of prime Illinois farm ground. A lane, which was what my grandma always called the drive way, was a tear drop shaped, white gravel affair that at its wide end circled the house. The gravel made a particular grinding sound when it was driven over that always made me feel welcomed. When the house was completed, Granddad learned, from his leathery wife, Jessie, that she had no intentions of living out on the prairie, even though the farm was a mere two-miles from town. As a consequence, Granddad build an identical house in Milford where Jessie could hold court and deem herself the center of Milford social life.

Since Great-Grandma Jessie deemed the house unacceptable, my grandparents took occupancy. Over time four boys and three girls were raised there, and I spent any number of summers haunting the barn, silo, and crib. I also helped to care for the chickens, cattle, and the occasional pig that were raised for food. Two of my uncles had served in World War II. Both of them had been grievously wounded and had returned to the homestead to recuperate. Uncle Don had been wounded by antiaircraft fire over Romania, and Uncle Sam had lost a kidney and several ribs on the island of Iwo Jima. They never discussed their service, and they especially did not talk about their injuries. Neither did my grandma. She had fully expected her children to survive, and she made sure that their recovery was as complete as medicine and a strong will would allow.

Out in the southeast corner of the large house lot, segregated from the pasture and the barnyard by wire fence, my grandma operated an extensive garden. As summer progressed toward fall, I would be recruited to help pick tomatoes, corn, squash, cucumbers, onions, carrots, and various herbs. The kitchen would become overwhelmed with the aromas of all these vegetables as they were cooked and canned for food during the winter. I never understood the term "canning" because the process placed the vegetables into quart sized glass jars. Nonetheless canning is what it was called, and it was what we did nonstop for several weeks in August.

There were expeditions to the woods that ran east of the farm. On these expeditions, we would harvest blackberries, raspberries, and the occasional stalks of rhubarb. All of this bounty was reduced in pots that never seemed to leave the stove. Once it was cooked, it was canned as jelly. Paraffin was melted and poured across the tops of the small glass jars.

A large, dense mass of rose bushes cordoned off the south side of the lawn from the farm ground. All forms of wild life took refuge in the thicket, most especially a pair of wood threshes whose songs my grandma reveled in. When I was old enough to have a gun, I got a .22 rifle, J.C. Higgins. I was free to lay waste to all the pigeons, starlings, and sparrows that my heart desired, but under no circumstances was I to shoot the songbirds that populated the yard and its trees. Also off limits were Grandpa's purple martins. "They eat ten times their weight in bugs everyday," was his reminder to any and all of why they were so valuable.

Events—and my last weekend before deployment was definitely an event in my grandma's mind—were required attendance by all family members. Those children who lived near were required to attend along with their families. On certain events, such as Grandpa and Grandma's fiftieth anniversary, even those children living some distance were required to attend. On that occasion, the house was filled to overflowing. Tables were set up in nearly every room in the house, and there were picnic tables set up around the lawn. Your proximity to the dining room table, a huge oak plateau that had four leaves to it and could seat sixteen with elbow room, indicated your rank in the family. Children and spouses were granted seats at the table. Grandchildren of all stripes were seated at various tables around the house. One ignorant cousin consistently used the back stairs as his table and chair. It was his defiance of the order of things. John was always looking for ways to confound decorum.

Linda hovered in the background, allowing my family time with me, a very generous act on her part. My grandma had no such attitude. The weekend went by with little time spent outside her attention. My favorite foods were prepared, and memories of our past were fondled like the rubbed smooth silver dollar that my grandpa carried in his pocket. Sunday came all too quickly, and it was time for us to leave. As I was packing, my grandpa appeared in the doorway of my bed room.

"Larry I want you to have this. Carry it for luck. " He extended his hand and revealed a silver dollar.

I took the gift and placed it in my pocket. "Thanks," was all I could muster.

"I've been carrying one for a long time, and I've always had good luck." He extended his hand, and as we shook hands, he said with a voice filled with too familiar emotions, "You're going to be just fine." He released the firm grip that always characterized his handshakes and turned away, leaving me to finish packing my clothes. I took a long look out the bedroom window. The porch roof that sloped away and where

I had spent many hot summer nights sleeping framed the lawn. Purple martins swooped across the lawn, vacuuming mosquitoes from the air. In the pasture, a half-dozen cows idly grazed on the grass indifferent to the warmth of the air or the drama taking place about them. Chickens wandered about the barnyard scratching at the ground and pecking at the uncovered riches. Realizing that this might be my last time seeing this vision, I found a new sense of wonder at how rich my life had been.

Linda was wonderful in her generosity. Never crowding, always there to spend what time we had to the best of her ability. There were a few times I managed to slip out of everyone's presence to get some things of a personal nature taken care of. One important item was the acquisition of a personal side arm. One of the instructors at Ft. Knox had advised, without equivocation, that an armor crewman needed a side arm. His observation was that the army did not completely agree with his assessment and thus did not issue every crewman a sidearm. His choice of weapon was the .45 caliber 1911, and he urged us to acquire one. In the middle of my second week of leave, I went about the task of getting a weapon. A gun store in Des Plaines had the very weapon I was seeking. I bought it, two magazines, and a holster that would slide on to a web belt. The salesman had a shoulder holster, but I decided that was too expensive and a little too flashy. I also purchased a utility knife as camouflage for the purpose of my shopping trip.

Having gotten what I needed to assure my survival, I returned home. When asked about my activities, I showed my sisters and Linda my newly purchased knife. Why I felt the need to hide the purchase of the pistol, I don't know; perhaps the pistol was a little too close to reality. Perhaps I felt it was a little too close to the violence I was soon to experience, and I didn't want to admit it into this quiet world that I had a few more days left to enjoy. I can't honestly say why.

Two days before I was to leave for Oakland, my Uncle Don and Aunt Marie took Mom, Linda, and me out to dinner at a very expensive restaurant in Elk Grove Village. When we had been seated and menus were handed out, the waiter asked what we all might like to drink. My uncle ordered his standard: a Scotch and soda. Aunt Marie ordered a Canadian Club and soda. My mom had a Canadian Club and Seven-Up. The waiter looked at me and asked what I would have. Before I could answer, my uncle offered that I would have a Scotch and soda too.

The waiter looked at me and immediately asked me for ID. Before I could admit to being too young to order my uncle intervened.

"He's in the army, and he'll be going overseas in a couple of days. He'll be fighting when he turns twenty-one. I think it would be alright to let it slide this one time, don't you?" The waiter looked around the table trapped in a conundrum: be the asshole that sticks to the rules or give in, break the rules, and risk being dismissed. Resignation took hold, and the look of "what the hell" took his face.

"And for the young lady?" he said looking at Linda. Resignation was in his voice. Having conceded the issue for me, there was no longer a defensible line to take.

"I'll have an old fashioned," Linda responded, acting as if she ordered drinks all the time.

"Very well." The waiter backed from the table and headed to the bar to obtain our cocktails.

When he was out of earshot, I leaned over toward my uncle and ventured that I would have been fine without getting a drink. His response was that I was going to be going into a very ugly situation and that the least that could be done for me was to get a good drink at a fine restaurant. Drinks came, and our orders were taken. Soon the conversation rolled to memories of Uncle Don, his service during World War II, and what life was like then.

We finished a fine meal and shared a cup of coffee. It seemed that none of us wanted to let go of the moment. At last my mom conceded the hour and suggested we needed to get going.

"Tomorrow will be a long day."

We were walking to the cars when Uncle Don took my hand. "Keep your head down. We are all expecting you to come home, young man." There was the tone of a command, not a suggestion, in his voice.

"Yes, sir. That's my plan, too." The handshake became a hug, and there was firm slap on my back.

"A year will pass faster than you know. We'll be keeping track." He turned and took Aunt Marie's arm, guiding her toward their car.

My mom, Linda, and I got into Mom's car and headed back to the house. The drive from Elk Grove Village to Arlington Hts. was a short one. It seemed funny to me that while I was growing up in Arlington Hts., every other community seemed a galaxy away. My world had been something of a set of isolation chambers. There was the home, Arlington Hts., the farm, Milford, and three visits to aunts and uncles in Texas and California. Everything else was trapped behind the glass screen of our television. I knew it was out there and it was real, but it was all distant. If the army had done nothing else, it had shrunk the world for me. Everything was now touchable, and if I could touch it, it could touch me.

A portion of the night was spent in conversation with Linda and Mom. Finally Mom gave up the ghost and went to bed, having the need to go in to work for at least a half a day. After we had talked about keeping in touch, I showed her a deck of cards. I gave her the two jokers and the box and explained that I would mail her a card from the deck each week and I would bring the queen of hearts home with me when my tour was done. It would be a good way of keeping track of the time, and it would encourage my diligent correspondence. Finally, we went to bed.

My sisters took leave of us early the next day, giving Linda and me our one final opportunity to spend time alone. My duffle bag was sitting in the kitchen, the pistol neatly packed toward the bottom between pairs of GI skivvies and inside a pad of GI socks. A thought had occurred to me as I was packing: Was it necessary to tell the airline that I had a gun in my duffle? I decided that it was not necessary. If it were found, I would plead ignorance of the requirement to tell anyone.

Once I had finished packing, I had gone into the living room and sat in the plush chair that occupied the corner of the house where two large picture windows met. I watched as a wall of dark foreboding clouds moved from the west. The once sunny day was now approaching the darkness of nighttime. Small gushes of wind stirred the leaves in the maple tree that dominated the front yard. There was one last gush of air, and then everything went still. Darkness grew so deep that you would have believed that the sun was just a rumor. Then there was the fall of the rain. It came in sheets, ricocheting off of the street and the driveway. In seconds there was so much water that a small stream formed in the curb and moved rapidly down the street.

I sat watching this display of awesome power thinking how many times I had watched this very same phenomena through these windows. Never in all those multiples had the occurrence so closely matched my mood. I could hear my sisters and Linda in the basement. They were discussing something that I could not clearly discern. Certainly it was not about my fate. I had seen and heard the stories of others who had gone and come back home crippled for life. That was not a fate I wished to endure. I made a pact with God, sitting there watching the storm wrought its might against the unyielding concrete and blacktop. Either I come home walking off of the airplane that brought me back, or I come back in a body bag. I wished no middle ground on the issue. Having satisfied myself that God and I were in accord, I returned to the observation of the storm.

In was not as dark now as it had been at the storm's apex. Cars were moving down the street with their headlights on, but this more out of caution than need. The next set of headlights veered into the driveway. Mom was home, and it was time to go. The backdoor squeaked its greeting, and the reedy voice of my mom called out. First for me and then for my sisters and Linda. Their responses echoed up the stairs from the basement. My mom's figure was outlined in the entry from the kitchen. The light from the ceiling gave her outline a glow.

"Lar, it's time to go." The tone of her voice was hollow.

We loaded into the Skylark and headed for the airport. The ride to O'Hare was a slow one. Traffic was backed up because of the thunderstorm. Several low spots in the road were covered with water, and cars were creeping through those spots backing traffic up. After looking at her watch for the hundredth time, my mom became more agitated. It was not a thing she wanted to be late for, but if I missed the plane, I would be with her for a little while longer.

At last we cleared off of the highway and turned onto the entrance to the O'Hare terminals. We circled around to the short-term parking and found a space. My mom wheeled and came to a stop. We all piled out and headed to the United ticket counter. We dodged traffic and rain dashing for the United terminal. The doors opened to a vast open hall that was fronted by a long counter. I joined a short line of patrons looking to purchase tickets for airports across the country. I bought a ticket for San Francisco and got the gate located in concourse B. My entourage made our way to the gate the flight would board from. In the flow and eddy of travelers headed to the four corners of the United States, we encountered Asycue, Radler, and Clary—three guys I had been in armor training with at Ft. Knox. Their flight was to Oakland, and it was to depart thirty minutes after mine. Since the three of them were landing in Oakland, we agreed to meet at the Oakland terminal. After we had settled on a time and place to rendezvous, I introduced my mom, sisters, and Linda. There were polite but curt greetings all around, and then Asycue made an excuse for his group to move on. As they moved toward their gate, Asycue turned and gave a low palm down goodbye swing of his right arm. I touched my forehead just above my right eye in a kind of salute, then turned to make my way to my gate.

When we got to the gate, there were about twenty people milling around. The boarding time was past, but no one looked like there was any interest in boarding. My mom opined in a half hope that perhaps the flight was cancelled because of the weather. A young guy looked up at us and advised that there was just a delay in boarding because of the storm, but they would be boarding soon. He looked at me in my uniform and knowing our destination surmised my final destination. "Sorry" was all he could add before plunging back into his book.

I felt like an empty rind with all the fruit hollowed out of me, and just a shell standing before some of the most important people of my life. I didn't want to think it, but it had a life of its own, and it was growing strong as time ticked on. Was this the last time I would see any of them? Before I had entered the army, and really up until I actually got orders for Viet Nam, time has seemed an infinite commodity to me. My family would always be around me. I would always have love and life to enjoy. Now time seemed an incredibly valuable thing indeed. I looked at all of them and opened for a hug. I told each that I loved them and that I would be back. A year is nothing. To Linda I reminded her of the deck of cards, and I handed her the ace of spades.

"That bad boy ain't coming with me. Not that I'm superstitious or any thing."

The young lady that was managing the gate called for boarding of the flight to San Francisco. I turned and headed for the jet way. The last thing I wanted was any more delay. If I waited another minute, I didn't think I would have the courage to go. I found a window seat easily as the plane was only carrying about seventy percent of its capacity. I took my place and pulled the shade up so I could watch the terminal for a final image of my family. There was none to be seen. One of the stewardesses closed the door, and almost immediately the plane lurched backwards. Passengers still trying to place carry-ons in the overhead grab for balance. The turbines began their whine as power was applied and the jet began to roll toward the taxi-way. The pilot announced our place in the take off line and assured us we would make our 9:30 scheduled arrival. I looked at my watch and saw it was just past 8:00. I could feel the big jet waddle toward the take off end of the runway. Every uneven spot on the jet way seemed to be amplified by the plane, making it waffle like a tight rope walker looking for balance.

At last the plane rotated onto the runway. The brakes groaned at the strain of stopping even this minute movement. Once the plane was aligned, the turbines picked up their pitch, and the brakes were released. The plane began its run as I watched the terminal race past the window. I could feel the rotation as the nose rose skyward looking for its true element. With a last hop, we were airborne and began to bank to the north. I watched the tidy grid of streets slip beneath the wing, and then it was all gone, obscured by the thick clouds. I could see flashes of lightening to the east out over where Lake Michigan would be. 'Not a good night to be out on the lake,' I thought to myself.

I put my head back against the back of the seat and was suddenly seized by exhaustion. It was a tired I hadn't felt since flight school. The next thing I knew, the pilot was announcing our impending arrival in San Francisco. Three hours had expired without my knowledge. As the pilot had promised, we touched down prior to the 9:30 ETA. Everyone rose to get their stowed personal items. I sat waiting for the aisle to clear. I didn't seem to be in any great hurry. A gentleman in an ash gray suit pulled his briefcase down from the overhead and looked at me smiling.

"Good luck to you, son," he offered as he turned and headed up the aisle.

I found my duffle riding the stainless steel carousel in baggage claim and headed for the taxi stand. I grabbed a cab and rode over the bridge to the Oakland Airport. In the dark, I could only see the lights that bordered the bay. It was really impossible to appreciate the bridge and its uniqueness. The cab pulled up to the terminal door of Eastern

Airlines, and the driver advised me of my charge. I had felt well off when I took off from O'Hare, but I had just blown a third of my money getting to Oakland. I paid the driver and got out of the cab. I had felt pretty comfortable in my ability to find my friends when we had arranged to rendezvous at the airport. Walking into the terminal, I was suddenly overwhelmed. There were soldiers everywhere. Some were in khaki, and others were in fatigues. They were all laughing and talking over each other.

I asked one guy in fatigues what the story was; were they being shipped to Viet Nam from the Oakland Airport?

"No, man, change of orders. We were all sent to Oakland to be shipped to the land of green, but there's been a cut back in men being sent to that pearl of the Orient. Old tricky Dick did what he promised, and we are all the beneficiaries."

"Where are you going? Where have you been assigned?" I asked.

He looked around the room at men variously sitting and standing in small clusters. "All over, man. Some are being sent home to await orders. Some are going to Germany. Some are being assigned to duty stations around the country. There's even a couple of poor bastards that are being assigned to Alaska. I guess it all depends on your MOS. Me, I'm going home to await further orders." He contemplated this last pronunciation and then added, "I heard that one guy got sent home to await orders and never got them. He spent the rest of his tour of duty rocking on granny's porch. I'd like to be that lucky!"

'Yeah,' I thought to myself. "Listen I've got to find some buddies that flew in tonight from Chicago. So good luck with that awaiting orders deal." I turned and headed through the crowd of men. I was just past a cluster of guys all going to Fort Dix. They all seemed particularly energized by their new assignment. One guy was assuring his friends that he had multiple female contacts in the area and he would be seeing to everyone's needs. I heard my name yelled and turned to see Asycue and Clary making their way toward me.

"Hey, man, weren't you listening? We were calling your name, and you just kept going."

"Sorry. It's hard to hear above this din," I offered as my apology.

"Well they have a right to be happy. I think we have dodged a bullet, gentlemen. Literally. No Viet Nam for us. And several guys that are going to Germany have armor MOSs, so I'm guessing that's where we are headed. Man, that's going to be one long ass flight. But the faultiness will help us forget that. "

I didn't know what to make of all of this. Thirty minutes earlier I had been contemplating life in the paddies and fields of a combat zone, dodging shells and Rags, and now I was to be shipped to Germany. I didn't know what to think. I followed Clary and Asycue, assuming that they were headed in the direction of Radler. We caught up to Radler at a hotdog stand eating a dog and fries.

"Jeez, Radler, can't you give your jaw a rest even for twenty minutes?" Asycue asked.

"What they served on that flight would not pass for dinner in most third world countries, let alone America."

"Right. Well I'm up for a little celebrating. What say we go over to San Francisco and party a little bit."

"I just spent a fortune driving over here from San Francisco. Now you want to go back?"

"Well when we made the plan, we weren't going to be assigned to Germany, and I wasn't much in the mood for a party. Now I'm going to German,. and I am ready to party! So let's load up and get the heck outta here." Asycue didn't wait for confirmation of his plan; he just headed to the door.

All of us dutifully followed Asycue through the doors and out to where taxis sat in a row waiting for arriving fares. In a blink, I was headed back across the bridge to San Francisco. When the driver asked for an address, Asycue replied anywhere we could get a drink. The driver nodded and continued over the bridge. "Here you go, gentlemen. If you can't find a drink in this three-block area, you ain't gonna find one," the driver advised as he pulled up to the curb. We clambered out of the cab and retrieved our duffels from the packed trunk.

Asycue leaned into the cab. "Where's a good cheap hotel?"

The driver laughed. "Around here they're all cheap. Good I don't know." He jutted his chin up the street. "The New Willard is up the street and around the corner." He laughed again and headed away from the curb.

"Well I wonder what that was all about? Screw it. Let's get a drink." Asycue picked up his duffle and started down the street.

"How about we get a couple of rooms and get rid of our bags then get a drink?" Clary asked.

"We stop at the first think we come to—hotel fine, bar fine. Okay?" Asycue responded with a tone of agitation in his voice.

"If we wheel into a bar, let's make sure they serve food!" Radler threw in.

"Jeez, you guys. A wet blanket has nothing on this crew. Let's just see what we can see." We fell in line behind a sailor and a soldier walking together. I found myself wondering what the deal was; why was a soldier hanging out with a squid? When the sailor bent forward to point to a spot across the street where a noise had just come from, a pair of handcuffs shown from the top of his trousers. The thing was he didn't have a Shore Patrol armband on. I pointed to what I'd just seen, and Clary responded in a voice loud enough for people in Oakland to hear, "Undercover Military Police. That ain't right man."

The sailor and the soldier gave us a glaring look and then crossed the street heading toward where the noise that had uncovered them had come. We came to the entrance to a bar, and Asycue made a right turn through the door, not waiting for discussion. We all dutifully followed him in. The bar was dark and smoke filled. There were two patrons at the bar sitting on stools that matched nothing else in the room. They were hunched in their posture, and it wouldn't have taken too much convincing to believe they were fixtures like the bar and the dirty mirror hanging on the wall. Their shirts were untucked in the back. In perfect unison, the two men turned their heads to see who had entered and then returned their gaze to the bartender.

The bartender, from what was visible over the bar, seemed a large, bulky man of average height. His face seemed to have been deprived of any attention for several days. He didn't even look up. There were several small tables of various geometric shapes with chairs placed around them in front and three booths that would accommodate a maximum of four people. The vinyl seats looked well worn, and a couple of cracks showed their age. And then all the way in the back corner, next to an unlit jukebox, was a large table with chairs scattered around it.

Looking at the furnishings, the first thing I thought was that you would have to seek out, on purpose, such complete mismatched furnishings. Literally there was not one stick of furniture that matched anything else in the room.

Clary said what I was thinking, "Hey, man, let's move on. I have bad feeling about this place."

"Don't look like they serve food here," Radler piped in.

"Sit down. This is gonna be fine. Just sit down, and we'll get a drink, and then we can head off to that hotel."

We took seats and moved them up to the table. Radler took the only captain's chair, and Asycue ended up with a chair with one short leg. We sat there patiently waiting for the bartender to take note of his new clientele. Finally Asycue had had enough of the waiting, and he slid his chair back, nearly flipping backwards because of the short leg. Mustering as much dignity as he could, he sauntered to the bar. After clearing his throat several times to no avail, Asycue asked directly about getting service.

"Let's see some ID," The bartender replied. He was clearly annoyed that his attempt to ignore us had not succeeded in discouraging us.

Asycue extracted his wallet from his hip pocket, removed a driver's license, and laid it on the bar. The bartender didn't even bother picking it up.

"Military ID. That picture on the driver's license don't even look like you, and you ain't no twenty-five years old." Asycue had tried to use his brother's expired driver's license. Asycue motioned for the rest of us to come forward. Like moths to a flame, we all walked forward to the bar. I handed mine first. "Close, but no cigar kid. Come back next month, and I'll help you out," the bartender offered handing me back the ID.

"I won't be here next month," I mumbled.

"Sorry about that." He looked at me closely and then added, "Next month you wouldn't be caught dead in a place like this. Not enough ambiance." The two men at the bar laughed at the comment. Each of us in our turn was denied service. Asycue tried one more time to get a drink, this time by advising that in a couple of days we'd be fighting for our country ,and it was a damn pity that we couldn't even get a drink before we went. "Well, good luck to you when you're over there, and when you come home, the first one is on the house." There was something in his tone that said he was being sincere. "You want a Coke? I'll give you one, but that's as far as she goes tonight."

We all declined and retrieving our duffels moved out into the street. The soldier sailor combination was headed our way, and we all decided that it was time to get off the street. As we walked along, Radler opined that it was bad karma to tell the bartender we were going to Viet Nam. It might undo the mojo we had going for us. Asycue pooh-poohed the notion and urged us to keep going. We moved at a brisk pace until we reached the corner where the New Willard Hotel was. The outside didn't look promising, but it was late and none of us had any reserves to keep going. This would have to do.

The lobby was a scared shell of what must have been an elegant lobby at one time. Gilding that had framed one wall was worn thin to nonexistence. Wallpaper that had, one would have to assume, been bright and elegant was now faded and stained. Seams were separating in several places. Paint was peeling, and there was a musky smell of old booze, cigar smoke, and cigarettes. If I hadn't been so broke and so tired, I would have turned around and walked out, never looking back and working hard to erase the images from my mind. But I was tired and broke, so I stayed with the other guys. We rented two rooms. Clary and I took one room, and Asycue and Radler took the other.

The elevator worked despite its appearance. We got the to the third floor and wended our way down a hallway that was barely lit by three low watt bulbs located at the two ends of the hallway. We found our rooms next to each other. We wished each other good night and went inside. Clary found the light switch and turned it. A *clack* loud enough to be heard in the lobby followed Clary's turning the switch that turned on the light in the ceiling. There was a momentary blink as the bulb came to life. The room was in much the same condition as the lobby. The décor, despite the room's segregation from general traffic, had faired no better. The years had been harsh. Clary poked the bed on both sides. "Which side you want? I don't think there's anything to be gained on either choice."

"Pick it, I don't care." I wasn't being magnanimous; I was truly indifferent. Clary pulled the bedspread back to reveal grayed sheets that ·hadn't been white since the beginning of Eisenhower's administration. "Wow. That's pretty bad."

"As long as there ain't any critters scurrying around, I don't care." Clary stripped off his trousers and shirt and hung them on the doorknob to the bathroom. "I'm gonna take care of some business you ok for a few?"

I looked at him and nodded that I had no urgent need for the facility. He pulled the door shut, and as it swung closed, it made a groan like the sound effect from some B movie horror pic. The nook that served as a closet without a door had a metal rod but no hangers, so I took my trousers and my shirt and hung them over the back of a wooden chair that sat in the corner of the room.

Clary came out and announced that the bathroom was all mine, such as it was. The floor and walls were all tiny white tiles with periodic black ones placed in no perceptible order or pattern. There were spots along the grout lines where dirt and miscellaneous grime had built up, changing the color of the grout from white to various shades of black and brown. A rust stain ran along a line from the curve of the wash basin to the drain, and a sturdy drip from the antique faucet followed the rust marked line to the drain. At first I thought that Clary had just not turned the handles tight enough, but after several attempts at quelling the flow of water, it was clear that the tap was not sealing completely.

Only one of the two light fixtures over the mirror had a working bulb, and the light in the ceiling had the lowest wattage bulb manufactured. The lack of lighting gave the room an eerie yellowish tinge. I brushed my teeth, and not wishing to leave the brush in this huge Petri dish, I took the brush and put it back in my duffle. Clary was on his side facing a wall that was in bad need of repair and paint. I could hear his soft, even breathing and assumed he was asleep. I turned out the light and fumbled my way back to the bed.

A mortar shell burst and searing pain followed by numbness shot through my left leg. I screamed for a medic to no avail. I felt for my calf and felt nothing. I screamed for a medic again, and still no one came.

I was about to yell out again when I felt Clary shaking me. "Wake up, man, you're having a bad dream." I sat up, and like in my dream, I could not feel my calf. My hand moved up and down the appendage. My leg was there, but it was numb to my touch. Then I realized that my leg had gone through a hole in the sheet. In my moving around in the night, the sheet had wrapped around my leg just below the knee, cutting off circulation like a tourniquet. Once I extracted my lower leg from the cotton boa constrictor feeling began to return to my leg. Soon I was feeling the pins and needles of circulation returning.

"Thanks man. That was weird."

"You telling me? I was in a nice place and all of a sudden I hear 'medic.' I wasn't ready for that, I can tell ya." Clary rolled his back to me. "What time is it anyway?"

I fumbled around for my watch. I turned on the light and saw it was a little after three. "A little after three."

"Great, I'm going back to sleep."

"Yeah," I responded. My lower leg was starting to smart with the return of blood to it. I rolled over and remembered nothing until I heard Clary in the bathroom.

Chapter Eight
DOWN THE RABBIT HOLE

"**F**uck me." There was the clatter of metal on tile and the sound of Clary shuffling around the bathroom.

"What are you doing in there?"

"It's about time you woke up. Get your ass in here and help me." There was the sound of metal clanking against the tile floor. "Never mind. I got 'em."

I rolled out of bed and stumbled to the bathroom. The metal waste can sat upside down in the corner of the bathroom opposite the door. A pile of towels lay across the floor. "What the hell are you doing?"

"I swear that is the biggest friggin' rat I have ever seen. If I had a saddle and some reins, I could ride that mother." A big smile seized his face at the fact that he had successfully trapped the mutant creature all by himself.

"Where?" I asked in my most skeptical voice.

"Under the can, man." He pointed at the inverted trashcan. Without thought I went toward the can to take a look. "Don't pick it up, dumbass; it'll get out." There was true terror in Clary's voice, and Clary was generally not much of a practical joker, so I backed up. "Let's just get cleaned up and get the heck outta here."

"Deal. Let me know when you're done." I heard the sound of scratching against the side of the can. Suddenly I wasn't that anxious to get into the bathroom. I retreated to the relative safety of the bedroom while Clary finished cleaning up. When he was done, I went in. I had intended on taking a shower, but hygiene suddenly didn't have that great an import. When we checked in, whoever was in charge could gig me for my lack of cleanliness all he wanted. I shaved and brushed my teeth, keeping one eye on the overturned can. As an afterthought, I picked up the towels on the floor and placed them on top as added weight to prevent the denizen's escape.

While I dressed, Clary went across the hall to see where Asycue and Radler were as far as getting ready was concerned. I could hear his voice growing more shrill as he related his combat heroics in the bathroom. Asycue expressed doubt and sounded like he was wanting to inspect the creature. Radler clearly had no interest. I had gotten my pants and shirt on and was pulling up one sock when Asycue appeared in the doorway. "Where's this Godzilla of the rodent world?"

"In the bathroom." I nodded toward the bathroom door. "If you're planning on looking, wait until I finish getting dressed. I wanna be out of here if that brut gets set free."

"What, you really think there's something under that trashcan?"

"I could hear it scratching at the side a little while ago. I don't much care how big it is; I don't want to be around when its free." I finished getting my socks and shoes on and packed up my duffle. As I headed out the door, I advised Asycue that it was all his. Clary edged past me through the door.

"Wait a second and let me get my stuff." He retrieved his duffle and backed out of the room. Asycue stood in the doorway a minute.

"You heard it scratching?"

"Swear to God." I raised my hand to the ceiling.

"Screw it. But I'm doubting it was that big."

"Whatever." Clary had no need to argue the point. And I, for my part, firmly believed that there had been a disproportionately large rat in our bathroom. I was willing to grant kudos to Clary for having single handedly trapped the creature. I had no requirement for more proof than the overturned trashcan and the scratching sounds on the side of the can. Anyone else needing more evidence was welcome to collect it as long as I was out of the way.

As we left the New Willard, Clary advised the man at the front desk of the creature waiting in his room. The look of doubt on the man's unshaven face told Clary that the maid who cleaned up our room was in for a surprise. Either her or the next unsuspecting guest. Radler was desperate for nutrition, and he started a campaign for us to stop somewhere to eat. His pleas went unrewarded as the rest of us were more interested in getting signed in.

"We get on the base and we'll get lunch for free," Asycue stated matter of factly.

"If I'm still alive then," Radler whined.

We were all signed in and assigned to a floor by 10:30. Once we had bedding and our floor assignment, we headed to our floor. The building we were housed in was comprised of huge open floors with bunk beds from wall to wall. Clearly the building was laid out to process large numbers of men with little concern for the amenities. We all found empty beds and made them up. We placed our duffels on the bed, and then Radler announced that he had officially hit a wall. If food was not available soon, he would expire.

We found the mess and got served. Radler hadn't even sat down long enough to get his bench warm before he had eaten everything on his tray.

"Damn, man, you better slow down, you're going to give yourself a belly ache," Clary admonished.

"Not to worry. I can handle the load." Radler got up and headed back into line. I picked at my tray. The food wasn't that appealing, and I had little appetite anyway. We were crowded in. We had been told as little as two weeks ago they were processing 1,500 men a week to Nam, and now they were cut back to 500. Finding slots for the overflow was taking time, and in the mean time men kept reporting in and having nowhere to go. We all assumed that we were going to be a part of that backlog.

Once we had eaten, we headed back to our floor. Of course the return was somewhat delayed by Radler's intake. Having recharged his batteries and staved off his eminent demise, Radler lead the way back to our floor. We spent the balance of the day absorbing intel from every imaginable resource. There were as many variables to our fate as there were men on the base. Asycue was a natural filter. Everything that he heard was processed to match his desired outcome—assignment to Germany.

Having mined every nugget of information, fact, fantasy, and pure fiction, we headed back to the mess hall for dinner. Apparently mining for information was as exhausting as walking around San Francisco had been the night before because Radler had an extraordinary appetite. By the time we had consumed our first service, Radler was clearing his third tray.

"Well I ain't sure where we are going, but I am sure it isn't Viet Nam," Asycue stated with the assurance of a first grade teacher explaining that two plus two was in fact four.

"You don't know that, and if you keep saying it you're going to jinx it for us," Radler opined between mouthfuls.

"You can't jinx this. It either is or it isn't, and everyone I'm talking to is saying the same thing. Old tricky Dick is doing what he said he would do. We are cutting back on manpower. Way back. Two thirds of the guys reporting here are going somewhere else. We're armor. The logical place for us is Germany facing down those Russian SOBs. And even if it isn't Germany, it will be Ft. Riley or Ft. Hood. What could be better that that?"

Despite Asycue's enthusiasm, the rest of us remained skeptical regarding our next assignment. I didn't think that Ft. Riley or Ft. Hood sounded all that inviting. Their only positive was that no one was shooting at you. Ft. Riley was the home of a military police training brigade, and I had heard that you couldn't wipe your nose without getting written up. The discussion went on for several hours without resolution. Finally, we all grew tired and decided to close the debate for the evening. It wasn't until I had showered and brushed my teeth that I realized that Wheelus wasn't here. I wondered to myself where he could be. Knowing him and his poor sense of time, he probably was just running late.

Conversation at breakfast started where it had left off the night before. One good tidbit was that donating blood would get you the day off from any duties. We finished our breakfast and headed over to the clinic to allow the vampires to draw a pint. I gave my pint and was examining the chit that had been given me. It didn't include the day that I was excused from duty. I basically had a blank check for goofing off. Satisfied that I would be coasting indefinitely, I headed back to my bunk. When I got to my floor, there stood Wheelus looking around for an empty bunk.

"Hey, man, you're late," I said.

"No kidding?" He smiled that wide grin of his. "I just had a sergeant down stairs explain that to me rather graphically. I think he offered me a rectal exam with his booted foot if I understood him correctly."

"So what's the deal? Where were you?"

"Canada, man. I was in Canada. It's beautiful up there."

"So why were you late? I mean, what were you thinking?"

"Canada? I was in Canada!" Wheelus had the look of a comedian waiting for the crowd to catch the punch line of what he considered a straightforward gag.

"Yeah, you said, but...." The bulb went on. I finally understood what he was trying to convey. "So why are you here?"

"Orders." He looked around and finally asked me. "Where is your bunk?"

I pointed to a bunk three in from the end and one up from the bottom. Wheelus spotted a bunk across from mine and threw his duffle and bedding on the empty mattress.

"Orders? If you were in Canada, why'd you come back? Orders doesn't explain that." I was mystified. Wheelus was a funny guy, quiet and introspective. He didn't chum around with many of the other guys in training. I liked him because he was never looking for an edge or trying to prove anything to anyone. I liked him even though he was a smoker of demon weed, which I avoided with vigor. I had never heard him argue for or against the war. In point of fact, I had never heard him express any opinions of a political nature.

"Well I pretty much had my mind made up to move up there. I don't see the need for all this bloodletting going on over there. I sure as hell don't want to do any myself. I am an American, and if my country says I have to go, then there you are. I didn't think that way going up, but I saw those beautiful redwoods and the coast line, and I didn't want to lose the opportunity to see those ever again." He chuckled more to himself than to me. "Thing is I probably won't ever go to see them when I get back. If I get back."

He started making up his cot. "Then there were you guys."

I was at a loss for what he meant. "What about us?"

"How could I go up there and live fat and happy while you guys went and fought?"

I shook my head. I didn't want his obligation to me to end up causing him harm. "Look don't lay any of this on me. If you wanted out, you should have stayed up there. I don't want you blaming me for this." I swept my hand to indicate the whole situation.

"You're off the hook. I ain't blaming anyone. It's all on me, so relax." He looked around the empty room. "So what's up?"

"Rumors! I have never seen anything like it. When we flew into Oakland, there was a mob of guys who had originally been scheduled to go to Viet Nam, but there has been a huge cut back in personnel going over. The hottest rumor is that we'll be reassigned to either Germany or Kansas or Texas."

"Wow! Looks like I made the right choice. I could have been stuck up in Canada for no reason." He paused and looked around the huge empty room. " Texas would be good. What do you think?"

"Germany makes more sense. Armor is the front line against the Soviets. Those M-551s are more practical in Germany." I was pleased with the strength of my logic. I had been leaning toward Germany without really putting too much thought into it. But now that I enunciated my thinking, it had the strength of logic to it.

"So when will they make the decision?"

"I don't imagine they'll keep us here too long. Maybe the end of the week."

"Think they'll send us home for another leave before we go over?"

"Nah. They got us all together; I think they'll just load us all up in a military transport and ship us all in one lump. Why send us all home and then have to pay for all of us to fly separately overseas?"

"Too bad. I wouldn't have minded seeing my girlfriend again."

"Ah well," was all I could muster. "I wouldn't mind seeing my fiancé again," I added as an after thought.

The rest of the day started in relaxation, but that status was short lived. When a work crew was formed, I employed my chit. The specialist looked at it and noting the absence of a date tried to get me in the crew. I displayed my puncture and argued that I was exempt. The specialist prevailed; I ended up in a group that was policing the lawns for trash and cigarette butts. My free pass had ended up being no pass at all.

We had policed the area to a Fair-thee-well with the energy of an overused battery. Clary spotted five other guys from our training unit in Ft. Knox who were headed to building 590; they had all signed in a day ahead of us. When he asked them what was with building 590, one of them said, "That's the rabbit hole, and it comes out in Than San Nout."

Our mood took a sharp turn. Interest in speculation about our ultimate assignment evaporated like spit on a hot griddle. All the rumors that had swirled around us now seemed more foolish than thoughtful. Asycue wondered aloud about all the guys we had seen at the Oakland Airport who were being reassigned. In answer to the question, one of the guys pointed out that they were all ground pounders. The 11 Bravos were the ones in over supply. The 11 Echos were the ones in demand, so we weren't going anywhere but the land of green across the pond.

In the late afternoon, after hours of quiet contemplation, we were all advised to rip up our bunks and bring the bedding along with our duffle bags to building 590. The building was a huge warehouse, open with bare support beams stretching up to the beams that ran from wall to wall. At the door, a bank of carts with canvas walls waited for the donations of soiled bedding. Temporary, in quality not duration of placement, partitions ran along the opposite wall hiding a stockpile of military gear for the soon to be departing troops. Once we were in the building, we all took sections of floor and sat down. Most of the guys sat along the walls, and as that territory filled up, the sea of OD moved in toward the center. Clary, Radler, and Asycue sat together, and without conscious choice, I found myself parked next to Wheelus. With as many men as there were in the building, there was an eerie quiet. Everyone knew where we were headed, but what we were to expect was a mystery as unfathomable as what lies at the bottom of Mariana Trench.

Wheelus drew his feet up to his butt and lied back with his head on his duffle. He gave all the appearance of a man waiting for his turn to purchase hot concert tickets. I felt it was just as well because I seemed to have lost my capacity for conversation. I looked over in the direction of Clary, Radler, and Asycue, and there didn't seem to be much interest in conversation there either. We were all headed to the land of green, and there didn't seem to be any paroles being offered up. Our destination was all that was on anyone's mind, and talking about it didn't feel appropriate right at the moment.

We had been sitting around for what seemed an eternity when we were instructed to line up. A sergeant stood over in a corner where the partition and the outside wall joined.

"You will form up starting here. At this first window, you will turn in everything but your personal items and your duffle bags. You will then proceed to the next window where you will be issued new clothing fitting for your new duty station." With the practiced precision of weeks of basic and advanced training, we formed up. If the army had succeeded in nothing else, it had at least taught all of us how to form a line. Slowly, with a steady pace that didn't seem to bog down at all, we reached the first window and turned in our class As, fatigues, underwear, boots, shoes, and our military socks.

Stripped of all our clothing, we each in turn moved to the next window. Here several GIs inquired as to size and then retrieved appropriate sets of fatigue pants and tops. The new fatigues were different from our standard issue fatigues. The blouse was loose, had multiple large flapped pockets, and was meant to be worn untucked. The trousers had cargo pockets and were lighter and fit more loosely than the fatigues we had been issued when we started basic. The jungle boots were a good distance from the leather pair we had been issued. The side panels were made of green nylon. The soles of the boots were a rubberized material that was supposed to be reinforced to prevent sharpened punji stakes from penetrating them. The nylon side panel was to allow the release of moisture and prevent various foot maladies associated with damp climates like those in South East Asia. They were also supposed to stand up better than leather.

I managed to give up my stuff and repack without drawing attention to my 1911. After we had changed into one of our sets of jungle fatigues and boots, there was a spontaneous outbreak of conversation as we all examined each other and felt our new clothing. It was not long before the quiet again took control of the open room as the newness of the clothes gave way to the realization of where these clothes were meant to be worn in a combat environment. Wheelus and I took our place, and he assumed his familiar pose.

It had been some time since the last of us in the room had been given his new fatigues and gotten dressed. With the idle time came a sudden restlessness. If we were going somewhere, I found myself wishing to get on with it. The quiet and inactivity allowed my mind to flow to multiple unpleasant thoughts. In a somber moment, I renewed my deal with God. Either I come home able to walk off the plane or I come home in a body bag. I had no taste for being crippled in any manner. It would be sometime later that it would occur to me that I could walk off an airplane without an arm or arms. I had just concluded my negotiations with God when we were told to line up at the door. A number of buses were waiting for us to board.

We exited the building and walked across an open space to a line of buses that sat with their diesels clanking in idle. The aisles were poorly lit, and walking down the narrow alley was a chore. The bus filled from front to back, and eventually the buses capacity was reached. I wondered what they would do if all the buses filled and there was one guy left over with no place to sit. In the dark of a 2:30 morning, the convoy wended its way through California's highways. I watched as a seemingly endless flow of vehicles streamed past us—all the varied occupants indifferent to our ultimate destination. How could there be so many people driving around at 2:30 in the morning? And, more importantly, how could they not take note of all these young men headed for combat on their behalf?

I had trained for this; I had believed in the justification for why we were in Viet Nam. I believed we were right to be there. But now, with the reality of it coming at me like a runaway bulldozer, my mind fought against the reality I was faced with. We passed through the gates of March Air Force Base and down past a row of hangers and parked airplanes. To me it looked like an inventory display of all types of Air Force aircraft. The bus finally pulled to a stop, and the door popped open like the lid to a jack-in-the box. We all filed down to the door and out into the sticky hot night. I had expected a C-141 to be our transport, and I looked around for one. My eyes settled on an aged D-8 with the words "Seaboard Airlines" stenciled along the fuselage over the passenger windows.

We placed our duffels on a conveyer that lifted them into the belly of the plane. We then walked around to the ramp that was pulled up alongside the plane at the front hatch. We all walked up the ramp and entered through the front hatch. A young stewardess greeted us and directed us to take our seats anywhere toward the rear. The forward part of the plane seemed reserved for officers and NCOs. Wheelus and I found a pair of seats toward the tail of the plane on the right side. It didn't take long for the entire plane to be filled. All had been done in silent efficiency with the whispered directions from the stewardesses.

Once the plane was full the stewardesses went through their safety spiel. There was a surreal quality to it all as we sat in our jungle fatigues awaiting delivery to a combat zone and as they described the emergency exits, emergency air masks, and flotation devices. The turbine engines began their windup, and the whine of the compressors filed the compartment. The ramp was moved away from the side of the plane, and slowly we began to roll. The groan of the brakes was a little unnerving when they were applied at the end of the taxi way. The plane pivoted, rolled, and pivoted again. We were in position for take off.

The take off roll was all too short. The nose rose up, and with a small hop we were airborne. I looked out the window wanting a last glimpse of home, any part of home. There was little to see in the blackness but the dots of light passing under the wings that were homes and cars and streetlights. I dropped my head back against the headrest of the seat. The captain came on to announce that our first stop would be Honolulu and that we would arrive there at about 6:00 A.M. local time. In Hawaii we would take on fuel and then head to Guam. The balance of the trip was left undescribed. We all knew what the final destination was. The stewardesses plied the aisles asking if anyone needed anything. The usual answer was no.

Ever so gradually the sun began to overtake our plane. Light, dim at first but growing, filled the sky. The ocean and the sky became separate entities with the increasing light. As it grew lighter, I realized that we must be nearing Hawaii since the pilot had said we would arrive around six. About the time I realized that we were nearing Hawaii, I felt the noticeable deceleration of the plane. The ocean below began to near, and there was nothing visible for it to land on. My ears popped, and still there was no land visible out my window. Wheelus took no note of the goings on as he awoke from sleep. The water grew larger out the window, and still there was nothing to land on. As the flaps were lowered and the airspeed lowered, the rocky land at last appeared beneath the right wing.

We taxied to the terminal and came to a stop. We were advised that while the plane was refueling, we could deplane and wander the terminal. Wheelus finally stirred himself and looked around. "Where are we?"

"Honolulu airport." I watched the bodies file down the aisle and then I added, "I'm going to take a stroll. You coming?"

"Yeah I guess." He stretched, and we both moved into the aisle behind the row of fatigue-clad bodies looking for an opportunity to walk around.

The terminal was small by O'Hare standards, and it was empty. There were shops, and they were open for business. There was a duty free store that several guys found alluring. Nugent, a fellow A-4-1 graduate, emerged with a beautiful Seiko watch. "Hundred and ten bucks," he announced. "In a store back home, it would be at least twice that."

Wheelus examined the purchase carefully. "Very nice, but I wouldn't have that where we're going. A year in that muck and humidity, that won't be anywhere near as nice."

Neugent looked at the watch and then at Wheelus. "It's quality so it will hold up where a Timex would die in the first week."

"Maybe, but I would send that baby home so it was waiting for me when I got there, and then you'd really have something."

"I bought it to wear, and that's what I'm going to do. It wouldn't be doing me any good sitting on a shelf in my bedroom." Neugent removed the watch from its case and put it on his wrist. A smile seized his face. I decided that Neugent was right. If you were going to spend that kind of money, you might as well get the use of it.

We walked to the end of the terminal. A set of glass doors welcomed travelers to paradise. A pair of MPs stood at ease, a baton held behind their backs. They watched us as we approached. "Where you headed, fellas?" one of them asked as we approached.

"Thought we might get some fresh air," Wheelus explained.

"There's plenty of that in the terminal," the MP replied without giving any physical emphasis to his comment.

We turned and headed in the direction we had come. "I guess they don't want us getting lost," Wheelus said after we were a good distance from the doors.

"Guess not," I confirmed.

We ended up laying over for two hours in Hawaii. After they had refueled the DC-8, it was discovered that one of the engines required maintenance. Wheelus was undisturbed by this turn of events. I wasn't particularly enthralled with the need for mechanical attention to the plane; it looked old and worn to me, and the need for attention on one of the engines just heightened my concerns. At last we were loaded back into the plane. Wheelus and I took our old seats, but this time he took the window. Seconds after we started our takeoff roll, land disappeared and the ocean again filled the window. We took a heading of west northwest, aiming for Guam.

To fill the time, Wheelus and I took up a game of honeymoon bridge. Looking back on it, I'm not sure how close to the real thing our game was. There was a lot of creativity employed in conjuring the rules. It didn't really matter; we played on not even keeping score, and we shuffled, dealt, bid, and played the hours away. Around noon we were served box lunches. Somewhere over the emptiness of the blue waters that filled the window, we went from Saturday to Sunday. A day lost with little note. After hours of playing cards, we lost interest and set the deck aside to catch some sleep. I don't know how much sleep I got, but it was interrupted by the plane again decelerating. The engines whined with their effort as the flaps were again extended and our altitude began to fall. I was better prepared this time for the lack of visible ground beneath our wings. The wings alternated height as the approach to the runway was adjusted. The engines alternated high and low RPMs as the pilots corrected their approach.

The tires chirped as they touched then recoiled before settling onto the runway. The engines roared as the air brakes were deployed, and then once the plane was slowed, they dropped to their low whine. The brakes groaned with their effort to bring the big plane to a stop. Before the plane came to a complete stop, the engines picked up again and the plane pivoted. As it rotated on its left wheel carriage, a panorama of an active Air Force base spun by the window. The plane came to a rest, and the pilot advised that we should debark since we would be here about two hours. A stair was wheeled up to the nose of the plane, and as soon as the stewardess announced it was permissible to do so, men began to unhook their seat belts and move into the aisle. Wheelus and I sat patiently waiting for the aisle to clear, and once the tail of the line was past the midway point ,we rose and began the slow side-to-side movement up the aisle.

The sun was bright, and the air was warm but not heavy with humidity. We found a strip of grass and took a seat. Across from our position, a grassy mound stood. It ran nearly the length of the concrete runway that sat, bleached white, in front of us. The huge vertical gray stabilizers of B-52s moved on the other side of the mound. To me they seemed like the fins of sharks cutting the water. As we watched, one by one the huge planes pivoted at the far end of the runway. Their enormous wings drooping to a point where a spindly wheel at the tip touched the ground.

Once the plane was in position, the eight engines began their chorus. The high whine was joined by a low roar, and then the big plane began to roll. As I watched the wing tips rise, almost independently, I found myself wondering if Wilbur and Orville ever imagined such a machine might evolve from their creation. By the time the plane reached our position, the air was filled with their roaring noise. I could see clusters of bombs under the wings. 'Someone was going to be receiving an invitation to hell in a few hours,' I thought. The wing tips were now flexed up above the top of the fuselage, and still the wheels stayed fixed to the runway.

At the end of the runway, the plane seemed to just roll off the edge and disappear. A layer of thick black smoke marked where the plane had disappeared. We all looked at each other in acknowledgement of having witnessed a tragedy of major proportions. It took a second to realize that we had not heard an explosion. Then someone pointed out to sea and said look. The B-52 trailing black smoke was flying a few hundred feet above the ocean slowly working for altitude.

"No friggin' way," someone in the group said. "Those guys have got to have a huge brass pair to do that shit every day!"

The whine and then roar drew our attention to the other end of the runway. Another B-52 sat poised for its takeoff run. The scene was repeated as the next plane rolled past us its wings demonstrating their remarkable flexibility. The Strategic Air Command logo painted on it nose and the star and stripes on its side drew testament to the plane's authors. Again and again until nine planes had taken off, the scene was repeated, and with each plane's introduction to its true medium, our admiration for the bravery of the crews was cemented.

At last our plane was ready. We all shepherded back up the ramp and to our seats. We could hear the turbines being fired up and then feel the roll of the plane back to the end of the runway. Unlike the B-52s laden with their payloads, the DC-8 found its element easily, and we were airborne before we reached the runway's end. We banked to the south, and those of us on the right side of the plane got a long last glimpse of the island of Guam, an unsinkable aircraft carrier in the middle of the Philippine Sea.

For a short time, Wheelus and I returned to our game of bridge. After an hour or so, we both lost interest. Wheelus snagged one of the stewardesses. He asked her about how quiet the plane was and if that was normal. The stewardess confirmed that the flights going over to Viet Nam were always very somber affairs. But, she added, the flights home were a completely different story. Everyone on those flights was cheery and forward. Wheelus opined that dealing with all those guys must be difficult. The stewardess said no, that was what made the whole trip worth it. She and the other stewardesses all appreciated what men returning home had endured, and if a pinch on the rump was all it took, then they were welcome to it.

When she left Wheelus turned to me. "That's worth fighting for. It's nice to know there are people who appreciate what we are about to do." All I could do was nod in agreement.

The flight was about two hours into this leg when we crossed over verdant forest. Everyone rose to take a look. None of us had expected to reach Viet Nam so quickly. The thick growth looked imposing. Thinking that soon that would be our home was very sobering. Shortly the forest disappeared, and ocean again filled the window.

"What the hell was that?" Wheelus asked no one in particular. He got no answer to his inquiry. I accepted the reprieve, however short it was, as a gift and rested my head against the back of the seat.

The deceleration of the plane alerted us all that we were about to arrive at our ultimate destination. We all began trying to catch a glimpse of the land below. The first thing I spotted after the all-encompassing jungle were a cluster of craters. Each had a small pond at its bottom. Then I saw a Huey sliding along the jungle ceiling. One rotor blade was painted with white stripes; the other was solid black. The color scheme gave the movement of the helicopter an undulating appearance.

Chapter Nine
THE WELCOME WAGON

We were losing altitude more rapidly than we had in our other approaches, and our air speed seemed greater. The plane touched down and began a rapid taxi from the end of the runway. We rolled to a stop, and the pilot announced that we were at Tan Son Nhut Air Base; the temperature was a balmy ninety-four, and the humidity was seventy percent. The plane taxied to a stop and the door in front was opened. The rush of humid air poured in overwhelming the air-conditioned interior of the plane. As soon as the stairs were rolled into place, we began the process of unloading. When the first man stepped onto the stairs, a roar went up from a mob of men sitting in the shade provided by a corrugated tin roof.

Someone yelled, "Gentlemen, our turtles are here. Took 'em long enough." Another yell went up as the new arrivals poured from the plane and started to assemble off to the side of the ramp.

A tanker truck had already rolled up under the wing, and a hose had been connected to begin the process of refueling the DC-8. I found myself wondering how I would react in twelve months when it was my turn to go home. Once everyone was off loaded, we were loaded onto deuce and a halfs for the ride to our next assignment. The road, such as it was, was a mire that put me in mind of peanut butter. A thick, reddish goo that clung to the wheels and sides of the truck. There was a steep embankment falling off on both sides, and people were making their way along the road.

Peasants with straw hats and sheer pants and shirts trudged along in the mud. They kept their heads down, purposely avoiding looking at any of us now invaders. Some of the pedestrians carried loads of a variety of descriptions. Across the horizon was a line of trees marking where the jungle commenced. Between the road and the tree line was a matrix of rice paddies, each marked by a raised dike system.

I looked behind the truck and saw a deep dark blue with slips of white that filled the sky. I suddenly realized I had no orientation to rely on. I did not know what direction we were going, and I had no idea in what direction the storm that could be seen was in. I had never felt so helpless. The odor was pungent as it rose from the sides of the road. The deep greens covered the ground like a thick shag rug. We crawled along the road until we came to a compound that was framed by a high berm. Towers bracketed the gate, which was a wood framed affair with barbed wire crisscrossing the frame. We passed through the gates and followed the road around the perimeter to an enclosure that was the home of the 90th Replacement Battalion. The barracks here were different from anything I had experienced in the States. There were gaps between the sideboards of about two inches. Covering the gap was mosquito webbing. The ceiling was corrugated tin roofing. Light bulbs were screwed into outlets nailed to the bare rafters. Three ceiling fans spun lazily to move the air in the building. It was here that we would become acclimated to the Republic of South Viet Nam.

Acclimation began with a pair of MPs who announced they would be going through our duffels and our personal items looking for contraband items. All our money would be swapped for script. No US currency would be allowed to be kept. There was a list of other items that were also banned. When my turn came, I handed over the few dollars I had and received paper that had various military images printed on it.

"Any more money?" The sergeant asked. I had contemplated lying to them about the silver dollar that my grandfather had given me for luck. But I come to the conclusion that they would understand that it was a lucky piece and not really money.

"Just this lucky piece," I said showing him the 1898 silver dollar.

"It's money kid; it has to be turned in." He handed me a dollar in script.

He emptied my duffle onto the cot. He went through the clothes and my dopp kit. "What's this?" He asked holding up my 1911.

"My personal weapon," I said matter of factly.

"Contraband. The army will decide what weapons you will need. We'll take this." He flipped the pistol to his cohort.

"I paid for that!" I challenged.

"You'll get a receipt." He took out a pad and filled out the form. He handed it to me. "When it's your turn to go home, you show that and they will take care of you."

"I won't need it when I go home," I responded.

"I don't make the rules; I just follow them so don't give me any shit." After a half an hour, the two turned to leave with their confiscated items. "Welcome to Viet Nam, gentlemen," One of them offered over his shoulder.

Once they were gone, I couldn't restrain my outrage any longer. "They shouldn't have taken my stuff. I mean my granddad gave me that silver dollar for luck. And what's with them taking my pistol?"

"Well I'm betting those items won't ever see the inside of an official inventory," someone opined. It was small comfort to me given my loss. The silver dollar was especially irking. We spent several hours harping about our items being stolen.

Our complaints were interrupted by dinner. I wasn't sure what my expectations were for food, but I was really surprised that we were served a good hot meal. Salisbury steak and mashed potatoes was definitely a good surprise, as was the quart of milk that was passed out with each plate. We lingered in the mess hall watching the activity of men coming and going. At last we could find no solid reason to loiter any longer, so we headed back to out barracks.

We hadn't been back more than a few minutes when we were mustered out. We were placed in two lines, and a couple of guys at portable desks called us forward one at a time. There was a lot of speculation about what we were doing. When the first guys to talk to the two men came back past the line, they advised us laughing as they did so that we should be prepared because they were looking for reenlistments. My first thought was for the guys we would be talking to. Talk about a tough sell: new guys just got in country twenty-four hours ago, and you are trying to convince them to reenlist?

My turn came, and when the guy saw that I had been a WOC he told me I could get an assignment as a door gunner if I would re-up for three-year. I told him I really appreciated the offer, but I would take my chances with whatever assignment came up. He said he understood and wished me good luck. I walked back toward the barracks shaking my head as I walked. When I got back to the barracks, I was greeted by a new scam. There were two men who were selling bibles to the new guys. They were Korean, and they worked for the publisher. All Bibles came with a lifetime warranty if any of the binding came undone. Talk about the perfect audience for their product! Who would not want to buy a Bible when they were facing combat? Any insurance was worth buying, and if a hundred dollars got the big guy on your side, then what the hell. Only three of the guys in the barracks didn't buy one of the Bibles.

The last news of the day had been that we were going to the 11th Armored Cavalry Regiment. The story that accompanied the news was that E-Troop had been ambushed, and there were a large number of casualties. It sounded to us like we were headed for E-Troop. When lights went out, I lay on my cot, and I suddenly became aware that I had not heard the sound of a single round of gunfire. How was it possible that I was here in a combat zone, and I had not heard one gunshot? It was all too confusing. I had thought a lot about what this all would be like but; so far it was nothing like what I had imagined, even when I allowed my mind to explore the wildest possibilities.

In the morning, we went to breakfast. There was little conversation as we all contemplated the future. None of us had any clear picture of what being in the 11th meant as far as activity, combat, or life in the field. The story that E-Troop had been severely beaten in an ambush did little to reassure any of us about our futures. We loitered in the mess hall not wishing to venture anywhere else. Any place we would go would only bring us closer to the inevitable. At last we could procrastinate no longer. One rose with his empty tray, and the rest drawn by laws of attraction copied the move.

When we returned to the barracks, we were greeted by a sergeant who was more than a little put out by our absence. "Gentlemen, get your shit together. We have a plane to catch, and here in the land of green, the flights don't sit around waiting for people to show up, even class A funnugies like you mopes." We stood looking at him with no comprehension of what he was saying. "Move!" the sergeant roared in a voice that would be the envy of any drill sergeant anywhere.

We quickly grabbed our duffels, scrambling to put them back together after last night's rectal exam. We walked to a concrete runway where a C-7 Caribou sat with its rear ramp down. We were moved around to the ramp and told to get on board. Two rows of collapsible framed jump seats sat waiting for us. In the center, there was a pair of wooden pallets about four feet high sitting consuming a majority of the space within the plane's fuselage. We all grabbed seats, and before we could buckle up, the ramp was rising and an air force sergeant was moving up to the cockpit.

I got my safety belt buckled just as the plane started to roll. In the space of time it would take for a bowling ball to roll the length of an alley, we were airborne. The plane banked sharply to the right and then snapped level. The noise in the fuselage was deafening. Various parts vibrated to no particular beat and certainly not in any harmony. I could feel the air moving through the cargo bay—not a comforting sensation. We had been in the air for what seemed like minutes when the plane began to slow and make a decent. I could hear the landing gear extend and then the flaps. Like one of those carnival rides that fall suddenly, we started to fall giving the short-term impression of weightlessness.

The ground grabbed the plane. The engines revved to spin the reverse pitched props. There was the sensation of skidding, like braking on ice. The ramp went down, and the Air Force sergeant ordered us up. We were instructed to grab the nylon straps that were hooked to the pallets. Duffels over one shoulder and the nylon straps wrapped over our empty shoulder, we started pulling. The pallets came clear of the floor mounts and started rolling toward the ramp. As I stepped off the ramp, my foot sank past my ankles in red goo.

I was wondering how we would move the first pallet out of the way through this goo to unload the second pallet when the plane began to taxi forward. Once there was sufficient space, the second pallet rolled down the ramp, and the ramp began to elevate. As the props increased in RPMs and pitch small globs of red clingy goo, propelled by the props, which began pelting all of us. I took a lead from one of the men on the ground and ducked behind the first pallet. The Caribou cleared the ground with a steep climb and a roll to the right. A sergeant ordered us to start helping to load the just delivered crates of ammunition onto a duce-and-a-half that sat to the side of the mud strip. I watched as the Caribou shrank in size then disappeared into the cloudy sky.

Chapter Ten
THE BLACKHORSE

Blackhorse base camp was an imposing affair. My first thought was that I was glad I would be on the inside defending and not on the outside trying to get in. The gate we passed through was another large wood framed affair with barbed wire lattice. A sign hung on the center of one side of the gates. On it was the adage of the 11[th:] "FIND THE BASTARDS AND THEN PILE ON!" I had to smile despite my growing terror at being in the war. The perimeter was set back one hundred yards and defined by a thick treeline. Everything between the wall of green and the foot of a high berm was plowed under foliage, stubble, mines, and trip flares. Strands of concertina wire were coiled along the perimeter about forty yards out. Rusty cans dangled from the tops of some of the coils. These cans were filled with ball bearings, stones, and anything else the might make noise if rattled.

A red road of thick goo wound off from the gate, disappearing into the jungle. The tops of towers stood over the berm. They were perches for monitoring the empty perimeter. We walked through the gate following the duce-and-a-half loaded with ammunition. I was surprised to see several Vietnamese walking around the inside of the compound defined by the berm's inside foot. Nothing seemed to me to be like I had expected. I had been in the country for two days and had yet to hear one round fired. And now I was confronted by the appearance of locals inside a major fortification.

We were directed to a barracks where we were told to select a cot. Once we had rid ourselves of our personals, we were told to form up in front of the building. Once we were assembled, the sergeant led us to a work area. A pile of empty nylon bags sat next to a small mound of the red clay that seemed to be the only type of soil in this country. There was a group of kids sitting on the ground, apparently waiting our arrival.

"These kids will be filling sandbags with you. Do not embarrass the 11th by allowing them to out work you! When you get a bag filled, stack it over here." He pointed to a spot where a row of already filled bags sat.

I felt a smirk as I thought about these little kids trying to out work us. At first we set a pace that had the kids' output almost doubled. But as the humidity and the heat built, our output slowed. The kids working at a smooth even pace began to catch us and then to pass us. After an hour and a half of progress, the sergeant came back. "Looks like the kids are kicking your asses, boys." He looked at the two stacks, about even in number, and shook his head. "Take a break, ladies, you look spent."

As we sat smoking and talking, we watched the children. They were playing some game where they were picking up round balls of mud. One child would throw the small round lumps onto the ground, then another child would toss a round lump into the air and then pick up a ball. After retrieving the ball from the ground, he inverted his hand and caught the round ball he had thrown into the air. Then he would repeat the toss and pick up two balls. With each toss into the air, the number of balls retrieved grew. Several of the guys using mime and hand gestures asked if they could try. The children allowed them into their game. Howls of joy rose up from the kids each time one of us failed to get past two or three balls retrieved. I suddenly had an admiration for these kids. They were happy, hardworking, and open despite the war raging around them.

We finished the task of filling sandbags and went to lunch. Once we were at the mess hall, the sergeant addressed us before letting us go eat. "Those kids will work all day never wavering in their output. They get $0.95 a day for the sandbags they fill. In addition to getting more sandbags to fortify the perimeter and our other defenses, it is important for you gentlemen to get an appreciation for what these people are capable of. They work hard, and they are not stupid slopes despite what you have heard."

Sufficiently schooled, we went to eat. Surprising, given our surroundings and our situation, the conversation revolved around the game we had tried to play with the kids and how hard they could work. We finished our lunch, and then we were taken to the armory where we were all issued a M-16 and a bandoleer of ammunition; finally, I felt like I was in a combat zone. We cleaned our new weapons and were taken to the range to fire them and get them zeroed. Each of us fired off a clip of ammunition, getting the rifles zeroed. Once we were finished, the sergeant advised us that there was intel that indicated we might be attacked in the next couple of days. He told us that if we were assaulted, we funnugies would be put on the line. "When those little bastards come at us, it's all hands out. There are no non-participants." He added that intelligence is more wrong than right, but it was never wise to assume that nothing will be happening when the heads-up was given.

That alert sent my stomach into spasms. Here I was, and soon I might be facing actual combat. What would I do? How would I handle being shot at? Would I do my job and not try to crawl into a hole and hide? The uncertainty scared me as much as the thought of being wounded. I was certain that I was no hero. They wouldn't see me galloping in front of enemy machine guns to take out an enemy position, of that I was certain, and my cramping belly affirmed that idea.

As it happened, there was no assault on Blackhorse base camp that night. There was my first mad minute though. Sometime in the middle of the night, the perimeter erupted in cover fire out toward the forest wall. These random acts of violence were meant to surprise the enemy and make it more difficult for him to assemble for an assault on the camp. It also allowed the men to release their pent up aggression on the surrounding jungle. The thought of catching Charlie with his trousers down around his ankles was pleasing, but in the absence of that pleasant outcome, there was always the idea that a communist tree had fallen before the righteous muzzle of your M-60.

That first mad minute had come without warning to us funnugies, and there was a moment of concern for all of us. Once the minute of heavy fire had ceased, a specialist came by and noticing us all at the ready with our M-16s, assured us with an explanation of Mike-Mikes. We headed back to our cots. I flopped onto the cot and found myself unable to close my eyes. 'At least,' I thought to myself, 'I had finally heard some gunfire.' Welcome to the Blackhorse.

Two days into my assignment with the 11th ACR, it was clear to me that the unit had plenty of swagger. Everything we were told was framed in the outline that the Blackhorse was the baddest unit in the Nam. As one instructor put it, "We don't generally sneak up on anyone. The tracks you will be riding in are noisy. A deaf cat could hear us coming. So they don't tangle with us very often because they know what the outcome will be. That does not mean that every once in a while some A hole on their side isn't going to feel his Wheaties and try us on for size."

In the morning, we returned to filling sandbags. When someone moaned about all of the bags already filled and deployed around the camp, the sergeant assured us all that when the mortar shells start falling, you'll wish you had filled ten times as many bags as you have. At noon we stopped and went to lunch. After lunch we went out into the open perimeter. It was here that we got some rudimentary training in demolition and explosives. We set up a claymore mine and detonated it. The claymore is an evil little device. The front, clearly marked as the side to be facing the enemy, is a wall of small ball bearings. Behind them is a wafer of C-4 that when detonated sends the ball bearings flying out in a wedge shape to inflict pain and suffering upon any enemy unlucky enough to be within the kill area.

We also learned about C-4, an explosive that comes in sticks, like butter. The C-4 can be molded or shaped in an unlimited number of forms like a lump of clay. Our instructor showed us how to shape a charge to maximize the damage done. He also lit a small piece and then placed a canteen cup of water over it. The C-4 burned with intensity and soon had the water boiling.

"You can burn this stuff to heat a meal, but after it is lit, you don't want to strike it because that would be your last act."

We moved on to DET cord. It looked like thick white nylon rope, and it was wound like nylon cord around a spool. He unwound some and handed the loose end of it to one of us.

"Take that end and head over to the sapling standing there." The kid took hold of the end with his forefinger and thumb and started walking toward a scrawny tree that had clearly been run over but had not been uprooted. Its defiance to being torn asunder was a symbol of the relentless jungle. Once he reached the tree, he wrapped his end a couple of times around the battered trunk. When he let go, the DET cord fell to the ground causing a chorus of laughter. "So okay, you can tie this stuff in a knot around to hold it in place." The kid picked up the end and taking a little more tied a knot that stayed in place after he released his grip. The sergeant waved him back to where the class was standing. There was a wooden post in the ground a few feet in front of where we were standing. The sergeant cut the DET cord and took the new end. He tied it in a knot and placed a blasting cap on his end. He connected the blasting cap to a firing detonator. He ran the wires that connected the blasting cap to the detonator out some ways, and as he backed away, the class followed his lead.

"The beauty of DET cord is that it explodes instantaneously along its whole length. When I twist the handle of the detonator, the DET cord will blow up the two ends at the same instant." The sergeant yelled, "Fire in the hole!" three times and then twisted the short handle on the detonator. As advertised the explosion occurred along the whole length at the same instant cutting both the tree and the wooden post in half. "This is really handy when you're wanting to clear a landing area quickly." He looked at all of us and saw that we were all looking at the two wrecked items at either end of the demonstration. "It's also handy for connecting charges that you want to set off simultaneously."

Several of us were given the opportunity to deploy the DET cord and destroy small items that were scattered about the open perimeter. We completed our lesson in explosives with instruction on how to shape an explosive charge to get a better result. Our lesson learned, we were dismissed for the day.

Our training in explosives had ended at dinnertime. We headed for the mess hall, and our conversation revolved around the lesson that we had just been taught. Like all American boys, we had all had experience with firecrackers, and our attempts to come up with comparisons fell short. We all agreed that even an M-80 was a weak sister to the explosives we had just seen demonstrated. As Asycue put it: We had just left kindergarten and moved into college where explosives were concerned.

As we walked back to our barracks, I watched the sun slip below the treeline, and I saw how the colors of night deepened before everything went to ebony. The green of the trees and the red of the ground became richer in the growing dark. I found myself wanting to hold on to that light, to keep it a while longer because, as we had all been told, the night belonged to Charlie. I suddenly became aware of the fact that I was standing alone in the empty kill zone. I turned and hurried to catch up with my fellow funnugies.

I noticed during dinner that conversations had begun to devolve to topics that had carried us through meals back in the world. I could not keep my mind focused on the relative merits of the Camaro verses the Mustang, and of course the 'vette. My mind could not release from thoughts of all the harm that laid a few hundred yards away. We finished dinner and went back to the barracks. All of the barracks in the country had the same configuration. The walls were wood slats with fine mesh to keep out the mosquitoes. The ceiling was bare wood rafters with corrugated tin. In the center, there was a row of fans that whirled idly. I noticed at the base of each ceiling fan, a small green lizard squatted, clinging to the tin roof. The lizards made no movement for long periods, then a head would turn a tongue would shoot forth, and an itinerant insect would become part of the lizard's meal.

The next morning we went outside the berm again. We sat in some bleachers at the edge of the runway. A different sergeant began to speak to us about the enemy. It is too easy to dismiss the foe as stupid, lazy, and unsophisticated, he began. "Charlie is a dedicated foe. He endures a lot of hardship with the sole intent of driving you out of his country. They listen to their propaganda officers' spiel about how evil we are. He knows he is here for the duration, and the only way he gets to go home is to get your ass out of here.

"The enemy is methodical. He will work as hard as necessary to get the results he wants even when he cannot hope to see those results. They are not above suicide to achieve their ends. They will not show you mercy, so you will not show them any. You have to be vigilant at all times, because if you aren't, one day you might go to the latrine, and when you're done, Charlie will be handing you your toilet paper. They have multiple ways to kill you: booby traps, mines, bullets, mortars, and all sorts of little treats they can hide in food or drink.

"When we come into contact with the enemy, it will be because they want the contact. We travel on ACAVs that make a lot of noise. We aren't sneaking up on anybody. Unlike those little bastards who can sneak up your pant leg and shake your willy for you after you've pissed."

At this point the sergeant turned and pointed to a patch of ground. The clump of foliage rose from the surrounding red muck and a smiling mud cover face appeared.

"Gentlemen, this is Joe. He is a Chieu hoi, a man who has abandoned the wanton ways of our enemy. He has come to strike fear into the evil Yankees who infest his country. Once he has been properly fed and had the opportunity to scare the shit out of you white devils, he will go back to his former NVA unit and attempt to outright kill as many of you as he can.

"You see that post over there?" He held his arm up pointing to a distant metal post. "Joe started crawling towards us from that point. He made it to where he is standing without any of you seeing him. If he were in his fighting mode, the AK-47 he is holding would have shared death with all of you. If you gentlemen choose not to be watchful, that is what will happen to you."

After that lecture, we were shown a variety of booby traps and weapons that the enemy could and would deploy against us. We were warned about the enemy's use of civilians, men women, and children. An example was given of a civilian worker. She worked in the kitchen. She began to show up with a growing belly. Everyone thought she was pregnant, but it turned out she was trying to gradually get her stomach to appear the size of a full-term pregnant woman so she could smuggle a land mine into the compound. A feat she managed but got caught before she could get the mine planted. She got caught because one of the cooks noticed that her belly went way down in just a half an hour. When he asked where her baby was, the girl panicked and the mine was found. No one asked what happened to the girl.

I had been in the country nine days now. I had been through dumb-dumb school and had the expectation that I would be sent out to join my new unit. Another surprise: We were all shipped to Di An (pronounced Zee Ann). It turned out that the 11th was being relocated to Di An, and Blackhorse base camp was being turned over to Marvin. The night before, I had written my first letter back to Linda. This was where my lie began. I told her that I had been selected to man a tank that was assigned to security for the US Embassy in Saigon. I hoped that this lie would spare her, and the rest of my family, any undue concern about my safety. One of the bennies of being stationed in Viet Nam was that I could mail all the letters I wanted by just writing free in the upper right hand corner of the envelope.

Di An was different from Blackhorse base camp. There was an air base with cement runways. The roads were firmer, and huge ditches ran along the roads. Culverts of corrugated tin ran under the dirt bridges. The perimeter, wherever it was in relationship to our base, was out of sight. It was in short like an American base. We funnugies were put to work at making our corner of Di An into the new base for Blackhorse. We remodeled one building, turning it into a club on one end and an office on the other. Barracks were cleaned out and set for occupancy.

The duty seemed light, and I had time to go to the PX and do some shopping. Amazing there was everything one could imagine for sale. It was like an amalgam of JC Penney, Sears, and all of the car dealerships you ever heard of. Electronics were on display in glass counters. Booths were set up where all of the car companies could show you display books full of pictures of what next year's automobile models would look like. If you wanted one, they could set up automatic deductions from your pay and have the car delivered to your local dealership in time for your return to the world. I bought a radio and a tape recorder.

Trucks rolled in every day delivering material from Blackhorse base camp. We would unload the trucks and store the supplies. We built a bunker for our ammunition, and we constructed a shower. The work went smoothly, and it seemed we had sufficient time off to go to the EM club or to just loll. I noticed that pot was an easily accessible recreational drug. I had never had any taste for it, and my intellectual curiosity was so weak that I didn't even wish to try it experimentally. Most guys restrained their use to evenings when the duties of the day were done. There were a few who pushed it to the limits. If you had the bent, it was hard not to indulge since it was sold in packs like cigarettes. You could buy a pack of Marlboros and have twenty perfectly rolled joints.

The day after we finished the shower, it was announced that some of us were going to have to go out to the field. We had all assumed that we would be staying in because the 2nd Squadron was coming in for a stand down soon. The fact was that the troops were diminished by casualties, and those rotating home and those in the field needed bolstering.

Chapter Eleven
F Troop

From the moment it was announced that we were going out into the field, I had experienced a tightening in my gut. I was here and now I would be in the field where the real action was, and I was scared. There was no sense in denying it to myself. I wondered if being scared at the prospect of going into the field made me feel this way, what the heck would I do when I got out there?

In the morning, six of us loaded onto a Caribou for a short hop to An Loc. An Loc was a provincial capital with a significant installation. The Caribou made a much different approach to the landing strip here. The approach was not as steep nor was it as fast. When the wheels touched down, there wasn't the sensation of skidding, and after touchdown we took a gentle taxi off to the side of the runway. The ramp was lowered, and we deplaned to a scene of much activity. UH-1 Hueys and CH-47 Chinooks were coming and going. Other helicopters sat idle off to the side on various landing pads. I found myself thinking that I should have been flying one of those instead of waiting to go out into the bush. An OV-10 came screaming in at tree top level and then rose in a high arc slowly rolling as it climbed. All this I watched with regret and sadness.

We were shuffled off to a CH-47, affectionately called Shit hooks, which sat off to a side. It appeared that there were pallets already loaded, and a Spec-5 with helmet visor down stood on the ramp. "Let's go, guys, we're running late." We clambered aboard, and the turbines began to whine. I took a seat up front on the left side of the helicopter. Looking out the window, I could see the shadows of the two main rotor blades begin to turn and pick up speed. Dust began to swirl in eddies that the blades created. Once the rotors were at full RPM, the Chinook rose slowly from the ground. I slipped my left hand down and pretended I was twisting the throttle on the collective pitch while my right hand held to cyclic. Once the helicopter was in a hover at a sufficient height, it turned about fifteen degrees, and the nose dipped creating motion up and forward. I watched the shadows as we rose, and I could see a huge rubber bladder suspended beneath us. I had not noticed that when we had boarded.

Once we were clear of the concertina wire that defined the base perimeter, we began to pick up speed. I had to look in amazement at all this aircraft was hauling. We had been flying in a straight line for about fifteen minutes when I heard a loud thud. I looked over, and a few feet to my right there was a slick round hole. Another hole was in the ceiling. The Spec-5 walked forward to where I was sitting and looked at the hole. He shook his head and leaned toward me, and in a shout meant to overcome the noise of the helicopter he said, "I just got done having one of those patched back by the ramp. Assholes, they just can't help popping off a round when they see one of us." He walked back toward the ramp shaking his head. That was all fine and good, but if I had chosen to sit just a few feet further back, I would be having a bullet pulled out of my posterior. The Chinook pitched back with its nose rising. We came to a hover about twenty feet off the ground. Weeds and grass fluttered under the pressure created by the huge rotors.

Looking out the window I could see we were just outside a circle of ACAVs. The ACAV is an M-113, Armored Personnel Carrier that had been re-rigged to be used as an assault vehicle more than a form of transportation for troops. The rear end has a ramp nearly the area of the back end. The rear three quarters of the roof is an open hatch. A cupola sat in the center just aft of the front engine compartment. The track commander sits there with an M-2 mounted within a quarter inch thick steel ballistic shield. The hatch lid and a couple of metal sheets complete the shielding. In the front left corner is the driver's hatch. The driver is probably the most vulnerable of the crew as the track moves into combat. On either side of the large open top were mounted two M-60 machine guns. Like the track commander's M-2, these two machine guns had a heavy steel ballistic shield providing five feet of protection to the front. Treads provided the locomotion, and they ran from front to rear laterally. The road wheels were torsion bar suspension, which provided a relatively smooth ride. Being on tracks allowed the ACAV to move through mud and boggy ground with relative ease.

Still in a hover a few feet off the ground, I watched as a couple of men rush in and unhook the bladder. Free of the bladder, the Chinook skirted off to the side and sat down. Debris was flying around, driven by the still spinning rotors despite there being almost no pitch. We unloaded the pallets, and with the help of men already on the ground, the crates were loaded on ACAVs. Unburdened by its load, the Chinook rose, rotated, and headed back the way it had come. An M-578 recovery vehicle had the bladder suspended from its boom, and all of the supplies that had been carried out had disappeared into the ACAVs.

A man, older than almost everyone else in view, came over to where we were standing. I stood in a daze over all the efficient action that had swirled around us. "Gentlemen, I am Captain Bailey, the CO of F-Troop. Unfortunately, your ride came in late so there isn't a lot of time. I'm sorry that we don't have time for formalities, so get on board one of the ACAVs, and we will be heading out." With that said, he whirled and headed for the troop command track.

We followed his lead and headed for the circle of ACAVs. I climbed aboard F-31, and the others selected various vehicles based mostly on proximity. Without delay the circle uncoiled and headed out from the site in a line down a dirt road that ran through the jungle wall. We hadn't been on the road more than a minute when a rumble echoed up through the column of tracks. I felt the slap of a shock wave on my back. The pressure was like a gratuitous back slap from a friend but covering my whole back. I turned to look back and felt another slap, this one harder. Columns of smoke and dirt were rising in pairs like giant pistons rising out of the ground, and they were headed in our direction. To me they seemed to be gaining on us, and as each pair of dirt columns rose, the shock waves grew stronger and more menacing. Just when I was certain we would be consumed in a ball of dirt and debris, the column stopped, and the echo of explosions too terrible to consider faded.

I looked at the man sitting in the commander cupola. He sat looking straight forward, seemingly oblivious to what had just happened. "What the hell was that?" I yelled.

The commander looked at me. "B-52 strike. The dinks like to go through our sites after we leave. They look for anything we might have left behind. So every once in a while we call for an air strike to catch them in the act. Problem was that you guys got out here late so the strike came a little too close to our departure." He appeared completely indifferent to how close we had just come to oblivion.

"Why didn't you just call the strike off or delay it an hour or so?"

"Those B-52s took off for this attack four or five hours ago. You aren't going to call that off. And as far as delaying the raid goes, they can't just pull over and park for a bit until we're ready for them. It is what it is." He looked at me and smiled. "No worries, you'll get used to it in time." He looked ahead and then turned. "I'm Bob. That's Rob and Ed," he said pointing at the two wing gunners, "and Larry's our driver."

I extended my hand to shake his. "I'm Larry."

"Two Larrys and two Bobs, this might get difficult. Well we'll work it out." He turned his attention to the road. As we traveled down the road, the ACAV settled into a relaxing, rocking motion. After a while, the motion came to be hypnotic.

We came upon a small village. A row of white plaster walled buildings ran along the road for some two hundred yards. There were clear signs of a fight that had occurred. Bullet holes and gouges from explosions and bullet strikes marred the white walls of some of the buildings. Bob saw me staring at the damaged buildings. "Ambush. Thing is when we came near the Vill, we noticed there weren't any villagers so we knew something was up. There was some serious ass kicking that day on both sides, but since we weren't surprised, we got the better of it." He smiled then added, "Big body count."

I looked along the roadside and noticed groups of locals. Some waved; most were indifferent to our passing. The children seemed the most animated by our moving through their world. I thought about the kids we had worked alongside at Blackhorse base camp. The thought seized me: What kind of life do these kids have? I could not fathom what life must be like given the damage that had been done to their home and stores when Luke chose to take on the US Army in their village.

On the other side of the village, three-one broke down. It was clear that this was not the first time the track had had mechanical problems. A cable was run from one of the other tracks, and we finished the march being towed. By late afternoon we reached our objective, an abandoned base camp. Wrecked concertina wire lay in pieces around the perimeter. Inside the concertina wire there were two berms. The large exterior one encompassed the entire site. The berm and the concertina wire ran parallel to each other with a gap of some forty yards. It had several areas where gaps had been cut by explosions. The smaller interior berm seemed more in tact and had served as a firing pit for either small artillery or mortars.

A platoon of ARVN were already in place. They were occupying the center berm. Once we were in the perimeter each track took its turn under the fuel bladder suspended from the boom of the M-578.

When everyone was topped off, the third platoon moved back toward the perimeter and set off to a chosen site for an ambush. The track I was on—three-one—was to remain with the rest of the troop inside the abandoned base because of its mechanical problems. Once we had established the outside perimeter, the ARVN filled in gaps. Everyone on three-one quickly set about preparing the perimeter. I was sent out with Ed to post trip flares. He demonstrated how to deploy the flares and set them up. Once we were finished setting the perimeter, we returned to the track.

While we had been setting the trip flares, the others had set up a tarp to shelter the cots. One end of the tarp was attached to the side of the ACAV; the other was supported by two long poles. There was just room enough under the tarp for four cots. Bob advised that we would be hot-cotting it. The man coming off guard would occupy the cot just vacated by the man going up on guard. The perimeter set and the track ready for the evening, we set about eating dinner. Tonight it would be C-rations.

Because there were five of us on three-one, the guard would only be an hour and a half each. Bob took me to the side and advised me that good crewmates were the ones who woke up promptly and took their post without giving the one they were replacing a hassle. I nodded my understanding. He asked me if I understood the mad minutes. I told him I knew about them but had never actually participated in one.

"You'll get a call on the headset letting you know when it's going to go. Just jack back the Fifty and let her rip. One box only, and don't burn out the barrel. Also they'll be doing radio checks periodically through the night. When they call our track number, all you have to do is hit the mike key and break squelch twice. That lets them know you're awake. It also confirms that the radio is working."   Bob looked at me and saw what was, I am sure, a dumbfounded look. "Don't worry, you'll be fine. One more thing: We keep a poncho over the cupola. When you take your place, just crawl up into it. It keeps the getting wet to a minimum."

Bob said he would take first watch, Ed would be next, then it would be my turn. It seemed to me that I had just closed my eyes when I was awakened for guard. I crawled up into the cupola and messed around finding the helmet. My period of guard duty passed uneventfully except for my gut tensioning imagination that envisioned a NVA soldier lying just on the edge of the jungle, the sights on his AK-47 zeroed in on me. Periodically the command track would call each vehicle on the perimeter. Ostensibly this was a check on the proper operation of the radios, but in reality the idea was to make sure every one was awake. I listened on the radio and in the air as the first radio check went off. "Two one radio check, if you read me break squelch twice." There was a moment of separation, and then static filled the helmet. "Two-two if you read me break squelch twice." The requisite static clicks echoed across the perimeter. At last my turn came, and I had figured the procedure out. I keyed the radio twice.

When my turn at guard ended, I climbed down and woke Bob number two for his turn. When he was out of the cot, I climbed in and to my surprise found sleep again quickly. Morning came, summoned by the sound cots being folded and people finding a place to relieve the pressure on their bladders. Light pushed back the darkness as the shadows cast by the tree line receded. The encampment went from shadowy gray to a hazy white. We had C-rations for breakfast. In basic training, we had been warned about the enemy's ability to turn trash into weapons. The C-Ration cans could house a hand grenade whose pin had been pulled. A line attached to the armed grenade would then be stretched across a trail or a road, and when the string was tripped over, the grenade would come out of the can. The spoon would flip off, and the grenade would detonate. In dumb-dumb school, we had been lectured about the same thing. When I went looking for a place to bury my trash, it was pointed out to me by the other Larry that there were empty cans and trash everywhere. It wasn't like makeshift weaponry was in short supply. Point taken; I lobbed my waste over the berm.

After we had packed and retrieved our trip flares, we and 33 track were sent to An Loc for the mechanics there to see if they could repair three-one. An Loc was a large town. When I had been flown up two days prior, I had not seen any of the town. There was a fountain in the square, and there were large numbers of people moving about. It was nothing like I had expected. I was coming to the realization that my concepts surrounding this country and what went on here were nowhere near reality. I could see a market open on a street that branched off of the square housing the fountain. We passed through the main gate to the military compound and rolled down to the motor pool. We were cut loose to find entertainment while Larry and Bob stayed with the track.

I found myself in the PX in An Loc. It was not nearly as expansive as the one in Di An, but still it had quite a variety of products. I had been paid, and the large sum of money was burning a hole in my pocket. Another advantage to being in service in Viet Nam was that there were no taxes on my income. When everything is net, a paycheck can be pretty fat even for a Spec 4. I found a 35 mm Canon Pelix that was within my budget, as if I had a budget. One thing, I was poor at was managing my money. What I earned, I spent. I bought the camera, several rolls of film, and some prepaid envelopes that I could use to send my pictures in for developing. I would chronicle my year in Viet Nam in film.

I got back to the track in time to hear the mechanics pronounce three-one ready for combat. As we loaded up, Bob saw my camera. "That ain't gonna last long in this environment in that box. If you use a 50 cal can that will keep it dry and safe from most abuse." He saw the prepaid envelopes. "When you send your film in for developing, where are you going to have the finished pictures sent?"

"Home, I guess."

"Really? I thought I heard someone say you told your family you're guarding the embassy in Saigon," Bob said in a tone that cast doubt on my choice without accusation. "When they see those pictures you'll be taking, how are they going to reconcile them with where you say you are?"

I hadn't thought that part through. "Guess I'll have them sent back to me."

"That'll do, but you'll need to keep them somewhere where this humidity won't do them in. I'd suggest another 50 can." He looked around. "Your problem is that there's a limited amount of space, and you'll have three other guys who are sharing what there is, and we for sure aren't giving up space for the ammunition."

I looked around. The floor of the ACAV was solid with cans of 7.62 and .50 caliber ammunition boxes. Cots and the tarp were rolled and placed against the front of the ACAV. For a vehicle large enough to haul around eleven infantrymen, there wasn't an inch of loose space. I would have to learn to accommodate my crewmates as well as satisfy my needs. I was part of a team now. Ed noticed my conundrum and with practiced efficiency found a place for me to stow the camera and a future pictures box. "There you go. Now you'll have to take a few pictures of me to send home." I shook his extended hand and said we had a deal.

We departed An Loc and headed back to the troop encampment. We hadn't gone half way back when the engine died. Several attempts to get it restarted failed. Bob, after a short fit of pique punctuated with expletives of a variety and density seldom encountered in polite society, completed his expression of displeasure with a smack against the cupola with his helmet. The helmet swing had as little effect as the cursing, but I think Bob felt better for it.

Having sated his ire Bob directed Larry, Bob #2, and me into the woods on either side of the road. "Go in about fifty yards, hunker down, and watch for anyone coming our way. Ed and I are going to see if we can get this piece of shit working." The thought of going out into the jungle fifty yards set a cramping in my gut that felt for all the world like a punch. I took my M-16 and a bandolier of clips. I climbed down the side of the track and started into the underbrush. "Hey, you planning on fighting off the whole NVA army?" I heard Bob call after me.

"What?"

"You don't need a whole bandolier, guy. We aren't setting up a defensive position. You just need to keep a lone dink from surprising us. If you run up against more than that you just need to get back and alert us."

I got his meaning but didn't feel like running the risk of running short on bullets. I turned and headed in the direction I had started in. When I could just see the track through the bush, I knelt down and started scanning the area in an approximate 240 degree arc. I didn't need to be told to stay alert; every nerve ending was on edge. My ears and eyes probed the surroundings for any hint of danger. Nothing would sneak up on me. I could hear the sounds of metal clanking against metal and oaths of various intensity coming from behind me. The thought of what we might do if we could not get the track running ran through my mind. Bob had called to the command track to let them know we had broken down. He had asked for a one-hour window. If he didn't get the track running by then, the M578 would have to come out and get us.

I had been surveying the area under my responsibility with increasing intensity. Every minute that we sat unmoving brought us closer to the doom I was sure was headed our way. The high-pitched whistle of a bird drew my attention to my direct front. My stomach was knotted almost too tight to bear. Just when I thought I could not take another minute out alone facing whatever terror the jungle might cloak, I heard the sweet sounds of the ACAV's engine. I didn't have to be called; I stood up, and trying not to look too anxious, walked with purpose to the waiting vehicle.

We made it back to the troop perimeter. First platoon was preparing to go out on ambush patrol when we rolled in. We took up a position on the berm and set about readying ourselves for the night. Ed and I again set about putting out trip flares. When we got back, C-rations were already being passed out, and the cots had been set up under the tarp. We finished our meal and were talking amongst ourselves about life back in the world when we heard the high-pitched screed coming from one of the clusters of ARVNs. A man stood up and started walking toward another cluster of ARVN soldiers, an M-16 waving in his one arm while the finger of his free hand poked the air in the direction he had just come from. He was yelling and despite the attempts of his comrades, continued in the direction of the other group.

"Oh bullshit!" I heard a baritone voice say. It was three-seven, the platoon sergeant. He was a tall, barrel-chested man with a red handlebar moustache. He carried himself with an air of total self-confidence. "I am not gonna get killed by some idiot in a huff over a stupid insult." He cut off the progress of the soldier with the M-16. He thrust out his hand and looked over in the direction he had just come. "Joe, tell this little shit to give me his fucking rifle."

Another Vietnamese voice called out from where three-seven had been. The man with the M-16 began to argue with the voice, pointing in two different directions as he talked. Three-seven again told Joe to have the man surrender his weapon. The voice responded again advising the man to give up his weapon. The ARVN looked around for support in his complaint and found none. It was clear that none of his compatriots would take his side against the American. Muttering something in a lower voice, he handed over his M-16. Three-seven took the rifle and pulled the clip. He ejected the chambered round and then looked around. Spotting the officer in charge of the ARVN contingent, he went over and handed him the weapon. "Don't give him that back unless there is an attack. Savvy?"

The officer nodded and shepherded the soldier back to his original group, nodding in understanding as his charge continued to air his complaint, but in a calmer tone. Three-seven returned to his track, and the night seized the encampment as the tones of black grew deeper and the ground and the jungle edge became indistinguishable. My tour on guard this night was punctuated by a mad minute. I fired off a box of fifty caliber into the jungle to my front. I kept the burst short and managed to stretch out the one box. Red tracer rounds sliced into the jungle disappearing into the dense foliage. It was amazing how far the .50 caliber round could travel despite the thick foliage.

The next morning, I was switched over to three-seven's track. They were going to be going on an ambush patrol, and he only had three crewmen. We went out to sweep the road north of Xuan Loc for mines. It was a tedious task, and I quickly realized that the man using the minesweeper was an open target. He walked along with headphones on, deaf to the world and the dangers around him. His need to watch where he was sweeping blocked off his visual perception of his surroundings. This was duty I wanted nothing to do with. We swept the road and found nothing. Apparently the local VC liked to slip out onto the road every once in a while to plant mines. This was especially true the night after we had swept the road. We finish walking the road and then turned our attention to a rubber plantation. We cruised through it, and even to my untrained eye, it was clear we weren't looking too hard for enemy contact. It was late afternoon by the time we made it back to the encampment.

We ate a meal that had been flown out and topped off the ACAV. The hot meal was wonderful, but even better was the quart of milk each of us was rationed. I got my first package from Linda; it was a tape cassette. Once we were set, we headed back out to an open area along the road we had swept that morning. The eight vehicles, nine minus three-one, pulled into a circle. A young kid who everyone called Cherry Boy and I went out to set trip flares. The track was faced out across the open land toward the jungle. As we planted the trip flares, I discovered that Cherry Boy had just a month of seniority in the country over me.

When we got back to the perimeter defined by the eight tracks, we discovered the rest of the platoon had built a huge pile of wood. Soon there was a raging bonfire going in the center. A bottle was being passed about, and everyone was having a good time. My nervousness at being on an ambush patrol for the first time was amplified a hundred times by the fact that there was a fire in the center of our perimeter that would highlight our silhouettes for all to see. We finished setting everything up and then retired. I was set to take my turn at watch from 2:00 to 4:00. I climbed into my cot and put the just received tape into my cassette player. I got the earpiece in and started to listen to Linda detail events since my departure.

She was beginning an admonition for me to not visit the Momma Sans and Baby Sans that she had heard about when a pair of black silk pants went draping over my face. I pulled them off of me and looked over to my left. There, outlined by the dying light from the bonfire, I could see three-seven mounting a person. He proceeded to have intercourse with all the requisite grunting and groaning. If you made something like this up, no one would believe it. They finished their activity, and I turned to put my back to them. No one else seemed to take any notice of the activity. In the morning, we packed everything up, and three-seven managed to move the woman into the vehicle without anyone taking notice. I don't know if 36, the platoon lieutenant, knew of three-seven's dalliance. It didn't matter because it went uncommented on.

The vehicles of the third platoon had reached their end. The entire squadron was in need of a stand down to get things fixed and refresh the troops. Part of F-Troop stayed out in the field an extra day to cover for the delay in First squadron taking the field. Because three-one was broken down again, 36 towed us back to Di An. An inglorious way to leave the field. It took two days to make it back to Di An. By the time we limped into Di An, the rest of the troop had caught up to us. They had been attacked on the part of F-Troop that had stayed in the field to cover. It had been a sortie by a small cadre of sappers who had intended to welcome the First squadron back out to the field.

We held a huge celebration in the courtyard of our new compound. A large round, corrugated tub filled with beer soda and ice in a mound that stood at least seven feet high. Everyone gathered around the huge lode of beverages. Soon everyone was imbibing a variety of beverages. Despite the huge pile of drinks, a liquor run was made necessary after three hours. Asycue came over to me after the celebration had been in full regalia for some hours. He had stayed out on Captain Bailey's track. He had spotted a sapper crawling up toward the command track and had tried to point him out to the captain. When the captain couldn't see him, he told Asycue to take the shot. "I didn't know that an M-16 could bounce a man, but it did."

As he related the story, I noticed his eyes; they were locked on something ten klicks behind me. His voice was flat, without inflection. As I listened to his story, I found a new thing to fear. I didn't want to be like that. I didn't want to go vacant over the act of taking another's life. If I had to I would take another soldier's life, but I did not want to be like my friend if it happened.

Chapter Twelve
THE STAND DOWN

The next day F-Troop started its job of rehabbing the tacks and ourselves. I got the bad news I was to be the driver on three-seven's track. I discovered why everyone called Cherry Boy by his nickname. He talked endlessly about sex. He was a virgin, and he could not have a conversation without bringing in sex. I didn't get it. If he was that obsessed, it was everywhere. Why didn't he just take out five bucks and get it over with? He seemed unwilling to do that, and he was equally unwilling to allow the subject to die a natural death.

Cherry Boy also had a habit of hanging on three-seven like a pet ferret. "Hey Sarge, you want this? Hey Sarge, can I get you a beer? Hey Sarge, what would you like me to do next?" It was all too annoying, and I developed a growing desire to be anywhere he wasn't. We spent the first morning stripping out our ACAV. We took all of our personal items off and stored them by our cots in the barracks. The ammunition was removed and stored in the newly built bunker. I had assumed that we would start cleaning the track and getting it ready for actual mechanical work when three-seven advised that the track was being turned over to three-one's crew; we were getting a new track, and the three-one crew would be cleaning up our old ACAV.

There were benefits to being in Di An that outweighed the hassles of details and formations. First and foremost was the absence of guard duty. We all could get eight hours of sleep every night. There was no need to climb into the cupola and watch with frightened eyes for a lone sapper or for massing assault troops. The second big advantage was sleeping under a permanent roof during the monsoon. Every night the sky would turn loose with a relentless torrent. The ditches on each side of all of the streets in Di An would fill to overflowing. The patter of huge rain drops off of the corrugated metal roof filled the big open room making conversation difficult. There was a comforting effect from the dark and the rain that made sleep come easily.

I could also lie on my bunk and watch the ceiling. The lights drew insects in Viet Nam just like they did in America. The difference here was that a platoon of lizards set up ambush patrols near each light. As I would lie, fascinated, the little creatures would set about eliminating nasty little bugs by the mouthful. They would barely move as their tongue would lash out with machine gun rapidity. No matter how many bugs they would eliminate, still more would come in to flutter about the lamps.

A third benefit was a theater of sorts. Actually it was a tented area that was open on two sides. The projector sat in the middle of the seating area and projected against a white screen that absorbed as much as it reflected. Wheelus and I had decided to go see *Cool Hand Luke* one night. We met up after dinner, and Wheelus was buzzed. The rain was pouring down, and the ditches were already filled to capacity. As we headed toward the street across using the causeway that bridged the ditch, I missed it and stepped off into the ditch. I was instantly soaked standing in water up to my neck. Wheelus stopped and, laughing, asked me why I was swimming. I grumbled something unmemorable and using Wheelus' assisting hand clambered out of the ditch.

Wheelus asked me if I wanted to go change. Seeing the volume of water still pouring forth and feeling that the air was warm, I declined. We went ahead and walked to the theater. Unfortunately, the projectionist was in his cups, too, and he showed reel three before reel two. The whole thing became a debacle. Except that Wheelus in his state of weed induced intoxication toughly enjoyed the confusion. My enjoyment was not enhanced by my wet clothes and the chair that kept sinking into the ground made wet by all of that dripping water off of everyone's ponchos. My mood went even darker when I discovered that my cigarettes were ruined as well.

Wheelus and I ambled back to the barracks. The rain had stopped close to the end of the show so there was no longer a need to hurry. Wheelus offered me a cigarette. I looked at it, suspect of what might be in it. It was too dark for me to see what, precisely, I was being offered. When I waved the offer off Wheelus laughed, "I know you don't do that shit. It's just a cigarette, man!"

"Thanks." I took the cigarette and then discovered that my lighter was also useless due to the soaking it had taken.

Wheelus laughed again. "Need a couple of chest compressions to get you breathing too?"

"Nah, I got it." By the time I had my cigarette lit, we were back at the barracks. Without hesitation Wheelus followed me in smoking a joint and laughing. I laughed at his lack of concern. Wheelus was truly a free spirit, and that was only enhanced when he was smoking.

I looked at him as he took a pull on the joint he held between his thumb and forefinger. He noticed my expression of concern. "You know I'm really better on guard when I smoking. I get paranoid as hell and hyper vigilant. Plus I get hungry, and I ain't gonna go to sleep hungry." There was nothing for it but to shake my head. We were friends who just happened to have two different views on pot. If Wheelus was forced to find one good thing about Viet Nam, it would be the almost limitless supply of pot. He seemed to have little interest in alcohol and absolutely no interest in any of the hard drugs. Pot was what he liked, and he chose not to discard it out of hand.

As I said, drugs were easily obtainable in the land of green. That didn't mean that everyone was a dope fiend, nor did it mean that everyone used drugs. There was a high percentage that used marijuana on occasion. The army's attitude seemed to be that as long as people were doing it with restraint and it was not affecting combat performance, then it was best left alone. The attitude was not so liberal with harder drugs like heroin. The tolerance for those was nonexistent, and the higher ups put forth an effort to clamp down on its use and its availability.

We hadn't been with the troop for long when one of the troopers was busted for distribution and use. There were a group of guys sitting in the barracks one evening when the man who was up on charges started loudly complaining about his impending court-martial. He opined that there was a NARC in the unit, and since he had gotten busted just after a group of us came in and there was one new guy who didn't do dope, it had to be that guy. He said it looking directly at me.

"I tell you one thing; that rat motherfucker gonna be wakin' up with a grenade in his shorts. That's for damn sure." A big smile filled his face.

I lied in bed that night wide awake. I wasn't sure how I was going to make it through the next several weeks if I had to keep one eye open for the impending attack. The next morning, my worry in that area evaporated when MPs came and hauled off the dealer. Someone of the guys in the barracks the night before had heard the threat and had implemented actions to circumvent my being fragged.

We had been down for a week. All of the weapons had been cleaned, as had the tracks. Mechanics had been through all of the vehicles doing long past due upkeep on the ACAVs. Everyone managed a shower every night. We were starting to look like a fighting unit again. At the end of the week, a fire started in the ammunition bunker. Three-two and his driver were the heroes. The driver backed up to the bunker, and three-two let the ramp down. He went in and using a couple of fire extinguishers put the fire out. By some miracle, no one was hurt. As sure as everyone had been that the bunker would explode at any minute, I knew that I would never have been the one to have done what three-two did.

I was advised that I was also taking a field maintenance course with a couple of other F-Troop personnel. Apparently the idea was to have a Sheridan expert in each one of the platoons to handle field issues as they arose. Russell was from first platoon. He was big, broad shouldered, and he carried himself with a no nonsense demeanor. Dawson was from second platoon. He was slender and dark haired. He extended his hand and introduced himself with a warm smile: "Hi, I'm Mick." I liked him instantly. There are people who are likeable by their nature; Mick was one of these. Even Russell fell pray. During brakes in our training, he and Mick began to conspire. A driving range and pro shop grew from broad outline to plans that became increasingly detailed.

I sat and listened as Mick shared his love of golf, and Russell provided his love of detail and practicality. As I listened, I could see every detail as they laid out their vision. There were conversations on other subjects, cars, girls, family but always discourse lay anchored on the business plan that would welcome them home. They were as firm in their vision as I was adrift in mine. I allowed the fact that I had no plans or goals to completely disappear in the creation that I watched form between these two. It was a wonderful escape from my reality, and I reveled in it.

I gradually became aware of a similarity Mick had to my grandpa. There was the same calm demeanor. The sense that Mick was totally comfortable with who he was and what he wanted. It seemed clear to me that he had a deep love of family and that he honored his dad and uncle. He wasn't a rabid, flag-waving, super patriot, but he wasn't ashamed of his country or his service. There was a lot to like in Mick, and I suspected that he and Russell would be hugely successful.

I was the only one of the three of us who had gone through Sheridan training, and I was the only Sheridan crewman. That was a little fact that had no impact on our being in this training. We all completed the three-day seminar on the workings of diesel engines that ran the Sheridans without any of us observing this inconsistency. The tank was equipped with a turbo charged Detroit Diesel. A six V fifty-three T engine as we had learned at Ft. Knox. One of the biggest discoveries in this class was that changing out the transmission required a torque wrench, and there were virtually none to be had in the Republic of Viet Nam. Our instructor advised us that if the transmission went out, we would be faced with a never-ending series of breakdowns. "My advice, the transmission goes out, drop a grenade down the hatch, and blow the thing up."

When the classes were over, I made every attempt to stay in touch with Russell and Mick. They allowed me to sit and listen to their plans, completely indifferent to my presence. I didn't care that I was not a part of this civilian conspiracy. While I sat and listened, I could lose my lack of a plan for all of life that waited on the other side of this experience. Listening to them, I didn't need to think about me; I could plunge deep into their future.

After the training, eighteen of us were loaded onto a bus. We ended up at a supply depot. When we got to the back lot and got off the bus, we were confronted by row after row of M-551s. Now it was apparent why we had been through the extra training at Ft. Knox. I was sent out to select a tank from all the tanks sitting there waiting for deployment. Once I had selected a tank, we went about moving vehicles around to get my choice out from the middle of the mosaic of olive drab tracks. Everyone having made their selection pulled into a line to convoy back to Di An.

I had experienced few joys like the one I experienced driving through Saigon traffic. Vietnamese drivers on motor scooters, motorcycles, and bikes of endless variety sped in and out of lines of traffic. Traffic laws were either nonexistent or they were ignored with such total contempt as to be useless. Speed, lane changing, turning into and out of the flow of traffic were completely unregulated by any measure I could discern. And driving was done with a complete disregard for any possible negative consequences.

The motorists, if they could be called that, were completely indifferent to our convoy and the size of our vehicles. Despite the encouragement from three-seven in my intercom headset, I continuously broke to avoid collisions with the civilian traffic that filled the streets we had to navigate. Some of the sportier natives even took to weaving through our convoy, speeding up to move from back to front, dodging in between the line of new Sheridans, then slowing to allow us all to pass them. Once they were in the rear, they would speed up and begin the twisting moving slalom run anew.

We finally got clear of the urban traffic and passed through the gates of Di An. Sanity returned as the roads cleared to just military vehicles that passed us. We turned into the lot where all of the troop vehicles were parked. I climbed out and went about stretching. The tension in my back and legs was unbearable. After arching and doing some toe touches, I contemplated going to get dinner. My decision to go to eat was overturned by three-seven who, for whatever reason, decided that I needed to go over the track thoroughly before I went to eat. Having assigned me to the task, he turned and headed for the mess hall, Cherry Boy in tow yipping like a pet Yorkie.

In a demonstration of obstinacy, I played with the settings of the seat and looked for personal storage space in the driver's compartment. Deciding that a sufficient amount of time had passed, I closed up the track and headed for dinner. When I got there, three-seven and Cherry Boy were finishing. I had a growing enmity for Cherry Boy and three-seven, I was sure, knew it. He did everything he could to roil the waters. As they left, I sat down and ate my meal in solitude.

The next day was spent modifying the Sheridans for Viet Nam. The fifty in the commander's cupola was mounted with a heavy metal faceplate. There were metal shields placed on both sides and a bustle rack was fitted around the rear of the turret. The numerals three and seven were painted on the side along with the crossed yellow sabers and numeral eleven. It took most of the day to get all of the Sheridans rigged.

We got back to the area late, and I was anticipating dinner when were ordered to go out. An M-578 recovery vehicle had broken down, and we were to go out and escort it back to Di An. It was growing dark, and my gut tightened at the thought of being out and about at night. Three-seven and three ACAVs were to go. As I was getting into the driver's seat, three-seven got into my face. "We have to get out there Mosh Scosh Riki Tiki. You go and you don't stop unless you hear me tell you to. Got that?"

My only thought was that I didn't need to be told that. I was going to go as fast as the Sheridan would allow. I didn't need him telling me. We formed up and headed out. The first leg was on paved roadway. The Sheridan hummed along without hesitation. We turned off on a dirt side road. The dark was deepening, and all I had were the slit lights to see where I was going. The road narrowed suddenly and fell off on both sides. Having only the slit lights helped me to keep my focus on the center of the road. We passed the narrow part, and the road widened again. I hadn't been able to see how far the ground fell off on either side, and I really didn't care. My only interest was in presenting a moving target that would be hard to hit if someone decided to take a shot at us.

We had been roaring down the road for some time. I had an image of the three tracks running behind me fighting through the mud I was throwing up. We came upon a roadblock of concertina wire strung across the road. I didn't hesitate, but I pushed down on the accelerator even harder. Soon sparks were flying from every direction as the wire wrapped itself within the pads of the tread. Wire was whipping up across the deck in front of me. Finally three-seven ordered me to stop.

"Damn, that's what I call a ride." He clambered down from the cupola as three Vietnamese came running down the slope from an emplacement I hadn't noticed before. Their arms were flailing, and they were wailing in their high-pitched voices, gesticulating and making a great show of anger. Three-seven seemed to take no notice of them as he started tugging at strands of wire trying to separate them from the tread. "What a frigging mess. Back up while I pull on this lose end." I eased the tank backward, and three-seven pulled; surprisingly the wire came free.

The ARVN soldiers continued to wail about the ruining of their roadblock, and three-seven persisted in ignoring them. Once he had the tank free from the wire, he pulled the mangled mess off to the side of the road toward the small compound. "There you little ass holes. You can straighten it out in the morning." He climbed back up into the cupola, and the next thing I heard was an order to move ahead slow; the recovery vehicle should be just ahead.

We didn't go very far when we saw the recovery vehicle sitting on the road with its engine cowl open and three men sitting on the side. "It's about time!" a voice boomed. "We been sitting here in the dark for a while all uncovered like a Playboy pin-up."

Three-seven was standing on the front deck to my right. "Oh boo-hoo. You girls a little nervous about the pajama wearin' boogey man?"

"Hey, we were all alone out here in the dark. All we had were three M-16s and a fifty. I was feeling a little naked, that's all. It wouldn't have taken a whole battalion to take us down, and that's not a good feeling." The man stopped suddenly in mid-complaint. "You the only track they sent?"

I looked back and saw that none of the other tracks were behind us. We were all alone with a broken vehicle. Where the hell were the others? I looked up at three-seven. "Where are the other tracks?"

"I don't know; they got hung up somewhere back there." He gestured over his shoulder. "We'll pick them up on the way back."

We got the cables and hooked up the recovery vehicle to our Sheridan and in a short time were headed back the way we had come. The concertina wire was no longer across the road when we came up to the ARVN checkpoint, and none of the soldiers showed themselves to us as we passed by. It took considerably longer going back, and it was almost half an hour before we reached the narrow bridge. The three ACAVs were on the other side, and two sat on the road across the land bridge; the other was on the embankment perilously close to flipping over. Several men were attempting to hook a cable to the rear of the track to pull it back up, but the single cable was too short to reach without putting the other track in jeopardy.

The recovery had a boom with a length of cable. We pulled the recovery vehicle across the bridge and hooked the cable to the rear of the ACAV. Amidst copious volumes of swearing and no small amount of sweat, the ACAV was pulled back up to the road. We had just gotten the ACAV unhooked when I was assaulted with a torrent of swearing and threats. "Friggin' maniac. Drivin' through the night like a strip-assed ape. You look back to see what happened to us?" said the driver of the ACAV teetering on the edge of the ravine. It seemed that all the other drivers held me to blame for their accident. I tried to defend myself, and I expected three-seven to shoulder some of the blame, but that was a false hope as he just stood back and let the three other drivers rip me up one side and down the other. At last he had heard enough and ordered everyone back into their vehicles. His only comment that even approached a defense of me was when he said, "Try and keep up ladies."

I felt a huge sense of relief when we passed through the gates of Di An and made our way back to our compound. We got the recovery vehicle unhooked and shut down the Sheridan. Three-seven clambered down from the turret and jumped to the ground. There was no comment made by him about the trip we had just made, just his backside walking toward the EM club. I was walking back to the barracks when the other three drivers followed me, yelping about my reckless driving and how I had almost managed to get them all killed by speeding over the bridge in the dark. My urge was to tell them all to screw themselves, but I was badly outnumbered and outsized. I opted to ignore their whines and grab a shower.

Once I had my towel in hand and it was clear where I was headed, the heckling ceased, and I was allowed my shower in peace. I stood in the stream of hot water and allowed it to erode my emotions as the water cut lines through the caked-on red dirt. This was one thing I would miss more than most: long hot showers in the security of a fortified perimeter. Di An had allowed me to turn twenty-one in relative calm. What the rest of the year had in store was impossible to guess.

Chapter Thirteen
BACK TO THE JUNGLE

On October 10[th,] our time in Di An came to an end. We started out to our assigned area around An Loc. The road march was where three-seven began his incessant diatribe about my tracking the track ahead of me. The idea being that staying in the tracks of the lead vehicle helped avoid the possibility of setting off a land mine. What seemed to be lost on three-seven was that the Sheridan was half again as heavy, and it was wider than the ACAVs that normally ran lead in any column. An ACAV could pass near a land mine, and even though I was nearly on that vehicle's footprint, I would still set off the buried mine because the Sheridan put greater PSI pressure on the ground, and that was what determined what set off most anti-vehicle mines.

Three-seven also made it clear as crystal that he would do nothing to assist me in driving the Sheridan. We were near mid-day on the march when we came on to a civilian convoy led by a bus followed by several motorcycles and a car. The road had narrowed with a grove of rubber trees to one side and a steep ravine on the other. The civilian vehicles had crowded up against an irrigation ditch that served the rubber trees. They had done this to give us room to pass. The bus was loaded to the ceiling, and a large collection of personal property was piled on the roof. There was so much jetsam pile upon the roof that it stuck out for several yards on the sides and back. The four lead ACAVs had passed the conglomeration without difficulty, but the Sheridans with its turret was too high to easily pass.

The turret was turned twenty degrees to the left pointing at the ravine. The bustle rack protruded out to the right. As I approached the bus, I could see that the bustle rack would snag on the miscellany loaded on the roof. At first I thought that three-seven was going help me by rotating the turret to allow the tank to get past the roadblock. Dead earphones in my helmet, he quickly made it clear that there was going to be no assistance. Once I had come up to the bus and the bustle rack was nearly touching the pile on the roof of the bus, I stopped. I had to stand as best I could while keeping my foot on the brake. I could see I was too close to the bus. I backed up, turned to the left, and inched back forward. I stood up again and could see I was close to the roof. I pivoted to the right, which allowed the bustle rack to pass without touching the belongings. I moved forward toward the rubber grove. I inched forward, leaving the road by several feet. Once I was completely clear, I pivoted back to drop back into line with the other vehicles. The other two Sheridans, with the assistance of their track commanders who turned their turrets to allow for passage without the pivot, fell into line behind me.

At the end of the first day, we made an ARVN compound near An Loc. This would be our NDP for the night. We set up our perimeter and prepared for the events to come. My gut was now in full shut down, and the discomfort was already setting in. The relentless rain creating a mire of everything not under foliage only added to the discomfort. A tarp attached on one side to the tank and set on poles at the other offered some shelter from the nightly onset of rain torrents. When my turn came to go on guard, the monsoon was in full-throated display. I had to put on a poncho, go out into the torrent, and climb up into the tank through the loader's hatch. Once in the turret, I took off my poncho and climbed up into the commander's cupola into the poncho that was left spread over the opening.

I was on guard for half an hour when the rains abated. On guard duty that night, I would swear every tree had a NVA soldier behind it. We had gotten a new toy while we were in Di An. The starlight scope could amplify any ambient light and in a green cast make the dark penetrable. I used it extensively, but it did little to calm my fears.

The next day we reached our operational center, FSB Marge. Located north of the village of Loc Nihn, it gave us the perfect site to conduct our daily activities. This would be our AO (area of operation) for some period of time. Our appearance was greeted with a flurry of capitalism. Older women and children from the village went about the job of selling locally bottled soda and cigarettes to us whenever we stopped along the roadsides. Young boys, their forefinger hooked into the first joint of their middle finger, would approach the men. The act was to symbolize their wares. Once they caught the eye of a soldier, they would go into their spiel about the availability of their virgin sister. Each boy was the head merchandiser of a large family of virgin hookers who all desperately desired sex with any, or all, of the troop. One of the guys, Mitchell, loathed these little hustlers. He called them the poon merchants.

There were other entrepreneurs. One young girl, who was very pretty, refused to have her picture taken unless she was paid in C-rations. Another young boy, always toting an M-1 carbine of World War II vintage, would exchange pictures of himself for packs of cigarettes. To my astonishment, he smoked the cigarettes even though he couldn't be more the ten or eleven years old. This gun totter also talked trash regarding the VC. He was forever offering to go with us on patrols in order to kill the VC.

In addition to selling, the locals were also involved in conducting banking. With every sale came an offer to "change twenty." They were forever looking to consolidate small change into larger denominations. This was necessary because the army would periodically recall all script and reissue new script making the old script worthless. This was a vain attempt to cull the black market. It had little effect. The locals were always in tune to these financial turns and having large denominations facilitated their ability to convert with the men in the field. They were remarkable in their network and how quickly they picked up on alterations to the financial landscape.

Our third day in the field was spent going on a med-cap to a Montagnard village. When we pulled in, our tracks were immediately surrounded by half-naked children all looking for any hand out available. Their teeth were stained various shades running from brown to black. When I asked what had caused it, I was told that they all chewed Betel Nuts and that was what discolored their mouths. Our medic, Doc Gizel, went to work examining the children and handing out antibiotics for several of the sick kids. I didn't have any candy to hand out, but I did have some note paper. I proceeded to fold a paper airplane which I launched to the delight of the kids. They all chased after it, and the one who got the plane returned it to me to launch again. Several other guys derided my efforts and offered that they could manufacture something much more flight worthy. I passed out paper, and soon several planes of various forms took flight, all to the squeals of delight of the kids who chased after them and then returned them to the soldier who had tossed them.

Captain Bailey had another mission. The village had been shelled the night before, and Captain Bailey was supposed to find out what had happened. He had begun an inspection when an ARVN officer from the compound that over looked the village came down. Captain Bailey had retrieved several pieces of the exploded mortar shells. He and the first sergeant were eyeing the pieces when the ARVN officer injected himself into the conversation. Captain Bailey called over Joe, our interpreter, to join the group. Soon a heated conversation erupted. I could hear Captain Bailey telling Joe to tell the ARVN office that the shelling was to stop and if the village was shelled again, his troop would return and flatten the ARVN compound to the ground.

The ARVN officer grew red and began gesticulating wildly, arms moving in all directions at once. Captain Bailey didn't back down; he just had Joe repeat his threat. The ARVN office stopped his rant in mid-sentence and turned abruptly to head back to his compound at the top of the hill. Doc finished his exams, and we all packed up to run back to the compound for the evening. We all took up positions on the perimeter and set trip flares for the night. After we had eaten, I went looking for one of the old hands to find out exactly what had happened.

Ed from my first ACAV was sitting on his cot drinking a soda. I walked up, and after some idle talk about how things were on three-seven track, I got to the point.

"What was that all about today? I thought it was just a med-cap, and it turned into a real dust up with that ARVN officer."

"The med-cap has only one reason for going to the Vill. They got shelled for a third time in a month. We were sent to investigate because there wasn't supposed to be any activity around that Vill at least not that frequently. The captain found frag from some of the mortar shells. They were US shells, not Chi-comm. So it had to be the ARVNs." Ed took a sip of Coke.

"Why would the ARVN shell their own village?" It made no sense to me.

"They hate the Montagnards as much as the NVA do. Hell, I heard that ARVN pilots, if they haven't gotten a mission or they couldn't hit their target, will drop ordinance on one of these vills just for fun." Ed took another sip of his soda.

"That's four kinds of fucked up!" I said reflexively.

"Those poor fools are the original men in the middle. I don't honestly know who they root for. The NVA don't usually shell a village; they just take the able men and women for labor. So these shellings are kind of unusual. That's why we are looking into it. I'd say the captain has pretty much put an end to those assholes shelling the Vill. But once we move out of the area, I don't know what will happen."

We finished our conversation covering mundane subjects like where we were likely to go next and Ed was going on R&R. I slid back to three-seven thinking about the horror that those kids in the village lived under and how lucky I was to live in America. I got back to the track in time to see three-seven pulling the radio mast down toward the rear deck of the tank. Cherry Boy was standing next to him, a couple of items of cloth in his hand. I watched from the ground as three-seven affixed a bra and then a pair of panties to the radio mast. My only thought was, 'Great, now I'll be killed in a track rigged with a set of ladies underwear. As if being loathed by my TC wasn't enough now I would have this public humiliation to deal with.'

My mind quickly jumped to the relief that would inevitably come when Captain Bailey saw the adornment and demanded it be removed. I was sure that this blight wouldn't last too long, but any time was more than long enough. Three-seven finished attaching the clothes and then released the mast. The bra and panties lifted into the air for all in the perimeter to see. Once this primary activity was complete, we set about putting up tarp and prepping the perimeter for the night.

One of the things I had not anticipated going out to the field was the frequent shuttling of hot food, ice, and drinks out to our NDPs. Every track had a cooler. When the Chinook would bring out supplies, ice and sodas were included. Also included with hot meals were quarts of whole milk. This was something I had never dreamed of and something that went a long way to easing duty in the field. We finished prep for the night, and I went and got a carton of milk. After everyone had been served, the remainder of the milk was opened for any who wanted more. I took a second quart. Waste not, want not.

About twice a month, sundries packs were brought out to the troop. Huge boxes filled with all sorts of necessities including TP, candy, and cigarettes were packaged for delivery. Camels were one of the cartons included. Everyone went for the Marlboro, Winston, Kool, and Salems. I saw an opening and an opportunity. I chose the Camels and got a full carton to myself. What I didn't smoke was available for barter.

The morning brought no expected relief. Three-six came by and expressed admiration for the new emblem. We ate breakfast, collected our trip flares, and set out on patrol. As we left the perimeter, we trooped by the command track where Captain Bailey watched us pass. There was a smile on his face. It wasn't explicit, but it was clear to me that he was smiling at the items dangling from our radio mast. Great, everyone was cool with this humiliation.

We ran a length of road north of our position and then swung west through an area of jungle that was not nearly as dense with a canopy that could be easily seen through. We fell on a clearing that had been spotted by a pink team the day before. There were old abandon bunkers and a pit where some cooking had occurred some time prior but not recently. There was nothing of interest after several of the men climbed down and went about probing the area. After half an hour, three-six called an end to the exploration. We loaded up and swung south to catch a road that ran through a rubber grove.

We broke out of the jungle on to the road. We were turning to make a circle back to our current base. We had traveled about a click when an explosion erupted toward the front of the column. I heard three-seven's voice directing me up the embankment and past the three ACAVs ahead of us in the column. Within seconds machine gun fire erupted from every vehicle in the column. When we came even with the front of the column, I could see three-two sitting with smoke rising from the engine compartment. Three-seven directed me to pivot and had me move forward into the grove.

I heard him direct Cherry Boy to load a canister round. The tank rocked backward and leaves, dirt, and other debris flew up in my face and into the driver's well. I suddenly realized that I was exposed and tried to swing the hatch shut. Once I had the hatch free, I spun it, but a HEAT shell that was stored next to me blocked the latch. Three-seven directed me to straddle an irrigation ditch that was two feet wide and about three feet deep. It ran through the grove and toward the jungle. We were about forty feet along the ditch when I saw a structure of logs and sand bags that created a bunker open at both ends. Something black was visible on the floor of the bunker.

No sooner had I spotted the black lump when a deep base thump echoed through the hatch. A stream of tracers flashed in front of me past the leading edge of the slope and into the black pile. I heard three-seven curse into the intercom at Cherry Boy, and then another stream of fifty caliber rounds flowed past me into the bunker. The reverberation and the roar of machine guns along the line were hammering my ears. All of third platoon was advancing in line through the grove toward the jungle that abutted the northern edge of the grove. Twenty yards short of the jungle wall, the advance halted, and three-six called a cease-fire. Short bursts of machine gun fire trailed off to silence.

"Get us off this ditch," three-seven's voice ordered.

"Where do you want to go?" I asked.

"You're the hot shot driver. You go whatever way you want to. Just get us off of the ditch."

To go forward ended in the jungle. Backing up without three-seven's guidance would be a nightmare. I took the bar and pulled left all the way to pivot. I hit the accelerator, and the tank began to pivot. Three-seven realized what I was about to do, and he started to order me to stop. It was too late; I had us rotating and now we were astride the ditch at a ninety-degree angle. I went forward, and we were now clear of the ditch. There was dead silence over the intercom.

Everyone clambered off the tracks and with personal weapons in hand proceeded to inspect what we had fallen upon. Several bunkers, like the one that three-seven had pelted, were arranged through out the grove. The ground was covered with leaves that rustled as we walked around the battle scene. Three-seven climbed into the bunker and exited with a bag that contained eight mortar shells. Off to the left, an NVA soldier's body was found.

Everyone headed over to see the enemy up close. When I got close enough, I could see black, silk like pants and top clad body. The right leg was nearly severed, and there were several huge gashes torn into the side and shoulder. The face carried the look of pain his wounds had inflicted before he died. Three-seven strode forward and grabbed the dead man's hair to lift the head up from the ground. Soon a picture was taken and then several other guys stepped forward to get their pictures taken too. I could feel a welling of stomach contents, and I fought the urge to throw up. The last thing I needed was for everyone to see that I had a weak stomach.

A search was begun of the site. Any intelligence or ordinance was looked for. Once an inventory of abandoned munitions and miscellany was complete, everyone started back to the vehicles. Three-seven and the LT were by three-two looking at the damage inflicted on the ACAV. The driver was sitting on the embankment, his heels drawn up to his butt, arms braced atop his knees. His forehead was resting on his arms. Three-two was looking at the hole in the engine cover and the shredded aluminum.

"This is bullshit. I get called on to lead out all the time. Look at that!" He was pointing at the ruined vehicle. "A couple of feet up or over and there would be more than enemy bodies to be added up."

"Quit your whining. You'll be getting a new engine and transmission. Nobody was hurt, were they? Someone has to lead, and the rest follow. That's the way it is." Three-seven clearly had no sympathy for Three-two or its crew.

I was standing in front of the ruined ACAV. What caught my attention was the proximity of the hole in the engine hatch to the driver's hatch. Two feet to the right as I faced the track and Bobby would be waiting for an evac chopper at the best or a body bag at the worst. The thought did nothing to settle my already churning stomach.

"I ain't whining. I don't mind being lead when it's my turn, but shit, I get lead all the time. I was lead the whole trek out of Di An. Then I'm lead today. That needs to stop."

"Well, you ain't gonna be lead now!" three-seven opined as he turned to look at three-two.

The lieutenant sniggered at this and turned to go back to his track. Three-seven waved at Rob on three-one. "Get around here and hitch him up, you can tow him back to the troop." Three-seven supervised the task of getting three-two placed in tow and then climbed up onto three-seven. I had already gotten into the driver's compartment awaiting orders to head out. "Alright take the lead. Follow this road to a T. When you get there, turn right and that will get us back to our NDP." We started out three-one now in the unfamiliar position of towing someone else and not being towed.

We made slow progress back to the compound. I hadn't realized how far we had traveled. We pulled in and went to the spot on the perimeter we had occupied the night before. Three-seven said nothing and, in point of fact, was away from the tank before I had it shut down. By the time I was out of the hatch, all I could see was him walking toward the command track and Cherry Boy trailing close behind. Rubin, the gunner, was sitting on the bustle rack wading through the equipment. He saw me looking at him and nodded. I climbed up to the turret top.

"What was three-seven swearing at Cherry Boy about back during that fire fight?"

Rubin smiled. Rubin didn't dislike me; in point of fact, my presence served as a shield focusing three-seven's temper on me and off of him. His attitude was one of go along to get along. He had little taste for Cherry Boy's constant ass kissing either. "After three-seven fired off that first canister round, the fire ring stuck in the breach block, and Cherry Boy couldn't reload another round. Pissed Sarge off something' awful "

The news brought a smile to my face. For a minute, I forgot the terror of the afternoon. Two days in the field and we had already had an engagement. It was not the severity of the contact, but the placement of that RPG next to the driver had my attention. If this was what the next ten months were going to be like, I would not make it. If the firefights didn't get me, three-seven and his attitude would.

The days began to pass steadily in a blur of regular duties, patrols, battle assessments, road sweeps, and occasional support for infantry actions. On one of these, I learned how unworldly the average gravel agitator was. One of the men riding on our Sheridan asked whether we could turn the main gun without turning the whole tank. When three-seven demonstrated the tank's capability, a "wow" rose from all those on board. Three-seven began spending more time away from our track. There were a number of patrols where two platoons worked together, and when third platoon was left in the NDP, three-seven would ride on one of the other tracks. I enjoyed these days the most. We had to stand guard on the perimeter, but I wasn't driving and I wasn't out looking for trouble.

Pink teams were two helicopters out on patrols. The Loach, Hughes OH-6 Cayuse, was the white part of the team. An AH-1 Cobra was the red part of the team. The Cobra would orbit at altitude while the loach would skip over the tree tops looking for activity. If he spotted something, or as more often happened, he drew fire, he would mark the area, and the Cobra would sweep in and devastate the site. As a follow up, we, or another army unit, would be sent to investigate the area more closely. Usually what we would discover were the remains of a camp sight. Abandoned bunkers, matted underbrush, and some disposable service items e.g., food wrappers, spent brass, and the occasional item of clothing.

One of the main duties was to patrol the roads. Keeping the arteries open was of prime importance. Our cruising of the roads was also an instigator of civilian activities. Travel and commerce depended on the roads being clear and unrestricted. When the monsoons stopped, the roads turned from mired goo to layers of dust. The dust became fine like talc. As we patrolled the roads, our bodies took on a hue, an orange-red coloration. It wasn't so much a tan as it was a stain that would not wash off. In the field, we all would shower under canvas bags filled with water. A sprinkler head allowed a flow that was barely sufficient to remove the soap lather let alone erode the layered powder. These attempts at good hygiene made little impact on the semi-permanent coloration.

We drove the roads looking for potential ambush points and to allow the locals to see us. To comfort themselves with the idea that we were there to protect them. We also swept the roads for mines. The local Viet Cong were adept at planting mines. They were also brazen. On one sweep that first platoon went on, they caught an old man and his two grandsons planting two mines. When he saw the ACAVs approaching, the old man saddled up his motor scooter and one grandchild and headed south. The abandoned child was caught. The platoon sergeant, Mendez, child in hand, asked the lieutenant what to do with the prisoner. The lieutenant said when he was caught being bad, a spanking was usually administered, and this seemed to be an activity that merited that level of punishment. Sgt. Mendez obliged with three quick swats to the child's posterior. Then Lieutenant had Joe tell the boy if his grandfather was caught planting mines again they would yank his chin bald. Thus the chided boy was sent packing in the direction his grandfather had fled, having never shed a tear. He also did not look back at us.

My fears of constant action against various sized units of enemy soldiers seemed foolish as time elapsed. The routine grew numbing as we moved from place to place looking for signs of activity and assuring locals of our presence to protect them. My conflict with three-seven was also more manageable as he spent increasing amounts of time with second platoon. I wasn't sure whether he was being groomed for advancement to master sergeant or second platoon's lieutenant needed the support of a seasoned veteran. Indeed a combination of the two was possible. Any way it was sliced up, he was away, and I didn't have to deal with his sarcasm and disdain.

The wisdom that had been imparted during Blackhorse base camp training was driven home as time elapsed. One of the sergeants had told us that the eleventh was not a stealth organization. We could be heard three clicks way, so any contact would have to be initiated by the NVA, and they only would initiate when they thought they had the edge. Everywhere we went we found evidence of NVA presence, past tense. The lack of firefights did not diminish my intestinal distress, but I did have some sense of ease as time passed. It was a most odd state of existence, but I accepted it.

Chapter Fourteen
THE FALL

The balance of October passed into November. The roads dried up, and thick red dust filled the air as we ground the drying mud into a fine powder with our treads. We were returning to the NDP after a road sweep. Three-seven was away with second platoon again, and Cherry Boy was riding in the command cupola. Dust was choking me, and my eyes burned from the irritation. The one advantage to having Cherry Boy in the command cupola was that he would keep an endless line of patter going over the intercom. He was in mid-regale about the virtues of various Italian singers when *WHAM*. An explosion filled the air. The force of the blast threw me up against the roof of the driver's hatch. For a moment everything was black. When I finally focused, Cherry Boy's voice filled my head set. He kept asking me if I was alright, and all I could think of was the thick dust that was choking and blinding me.

My first thought was that we had been hit by an RPG and a firefight was about to erupt with Cherry Boy in charge. When no responsive gunfire started, I realized that I had hit a mine. My head started to clear, and I told Cherry Boy I was fine. Then I climbed out of the hatch to inspect the damage. The right idler wheel and the first two road wheels were wrecked. There was a two-foot section of tread that was gone. A crater sat underneath the right front of the tank that was at least two foot deep.

Once everyone had an opportunity to inspect the damage and inquire as to my physical status, we set about the task of short tracking the right side so that we could get back under way. We ran the track over the drive wheel and looped it around the middle road wheel. I unbolted what was left of the idler wheel and the second road wheel. With some help, we extracted the torsion bars. We tested the tension on the shortened track, and all agreeing that it would hold, we headed back to the NDP.

When we got back, three-seven called in, and when he heard that I had hit a mine, his only thought or comment was, "Maybe he'll start doing a better job of tracking now." No inquiry about how I was or whether anyone was hurt. The company mechanic called for replacement parts and ran down the list. That night, between stints of guard duty, I removed all of the damaged parts still bolted to the hull. In the morning, the mechanic came by and said we were ready when the parts came in. The aluminum skin on the hull had been peeled back by the blast, and I wanted to cut it off, but there were no tin shears to get the job done. The best we could come up with was to bend it back and bolt it to itself.

In the morning, the rest of third platoon headed out on a recon of a rubber plantation that we had inspected a half-dozen time already. Their recon was cut short when a call came in that a Loach had been shot down in a clearing not too far to the south. The Cobra, trying to protect the crew of the Loach, had made an error in estimating the strength of the force they had just discovered. A Chinese .51 set at one end of a defensive line had brought him down as well. All three troops were sending parts of their units to the site.

Because three-seven was still damaged, we were left with first platoon to defend the NDP while everyone else set about rescuing the downed crews. From descriptions related to me the next day, it was good old-fashioned cavalry charge. E and G troops entered the clearing spreading left and right; as a gap opened in the middle, F-Troop rolled in. They charged to the tree line across the clearing where a battalion of NVA were dug in. Artillery and air strikes fell on the NVA's defensive line, and overwhelmed by the combined fire, the enemy had to surrender the field to a superior force. Two guys in E-Troop and a couple from G-Troop were wounded, but overall the NVA paid a horrible price.

Unfortunately, so did we. The Loach driver and the crewman were killed in the crash. Once the helicopter impacted, it caught fire. Some of the guys described the bodies of the men as being horribly burned. The crew of the Cobra fared better. Though wounded and their Cobra a total wreck, they were at least alive.

By the time it was over and the enemy had withdrawn, it was too late for the three units to withdraw. So they set a defensive perimeter and hunkered down for the night. Their problem was that they were all low on ammunition. During the night, they watched as lanterns moved in through the trees to extract their dead and wounded. First thing in the morning, two supply choppers and a dust off came out to the clearing. They extracted the bodies and the wounded while everyone from the three troops hustled to resupply themselves.

While the rest of the troop was engaged in the fight, we waited for the supply chopper to get to FSB Marge, our new home. When the Chinook with our parts arrived, we started working on repairing the damage. Once it was dark, we pulled the tarp up over the front of the tank and continued to work. The heat and the lack of fresh air made the job stifling. Sweat mixed with dust clogged our pores and made breathing a struggle. As morning approached, we finally managed to get the new length of tread on and test the tension. We were good to go just when everyone else returned. There was exuberance over our success. The official body count was fifty-three. The guesstimation was much higher. Smitty, a California lad built like a line backer, offered that he and the other wing man on three-four had accounted for fifty of the total body count, and artillery and air strikes got the other three.

I felt somehow cheated by being down while the biggest contact we were likely to make went without me participating. Now I would be back to dealing with three-seven and his behaviors. There was nothing for it but to hunker down and deal with it as best I could. I was sure I would get through it some how. I just had to close my mind to his attitude. I grabbed on to something my grandpa had said a long to ago: "I don't give a darn about the opinion of people I don't care about. Whoever they are, if I don't care about them then I sure as heck don't care about their opinion." I decided I would apply this to three-seven.

We were moving our operational area more and more northward along highway 13. We were also probing closer and closer to the Cambodian border. As we moved north, Captain Bailey set down a dictum that there would be no more fraternization with the locals, especially the working girls. This didn't go down well with three-seven. If there was anything he was into, it was sex—more so than even fighting and killing enemy troops. It had been several weeks since his last dalliance, and a working girl showed up at the perimeter. Seeing an opportunity, three-seven headed into some tall grass twenty yards outside the perimeter. Soon they disappeared within the cloak of the tall grass. Unfortunately for three-seven, this did not go unobserved.

Captain Bailey showed up behind our track and asked where three-seven was. I shrugged and was about to lie about knowing where he was when the captain said, "Never mind, just give me your M-79 and a tear gas round." I complied with this request. Captain Bailey gauged the range and lobbed the shell into the tall grass. A pop and a cloud of thick gas rose where he had aimed. Three-seven's head popped up. "Who's the smart ass?" he bellowed.

"That would be me, Sergeant." He turned and started walking, not waiting to see what was happening behind him. "In my track now!" The captain's path took him past where I was standing, and he handed me the M-79, continuing his trek to the command track sitting in the middle of the NDP. Three-seven came huffing past the track without looking at any of us hitching up his trousers as he did so. I couldn't help but smile as I saw him disappear into the command track, and the sound of Captain Bailey's voice rose to a level I had not heard since our encounter at the Montagnard village.

The dressing down took more than ten minutes. Three-seven's voice was never heard, only Captain Bailey expressing outrage at being defied so blatantly. The diatribe concluded with an admonition that three-seven was too good an NCO to let sex drag him down. He needed to heed Captain Bailey's directives and improve his personal attitudes. Three-seven exited the command track with a smile, that Captain Bailey could not see, filling his face and sending the handlebar ends of his mustache nearly reaching the corners of his eyes.

Unfortunately for three-seven, this was not the last incident of insubordination. He was again with second platoon, and Captain Bailey and first platoon were together on a recon of an encampment that had been attacked by a B-52 raid. The object was to assess the damage and see what could be discerned about size and location of the unit that had been bombed. The squadron commander was supervising the action from his helicopter and calling down with instructions. At some point in the operation, when he thought the colonel was out of radio communication three-seven made a wholly inappropriate comment. The colonel, still on the radio, heard this, and the op came to an abrupt end.

We were sitting in the perimeter when the colonel's helicopter landed. Within half an hour, the rest of the troop broke out of the treeline and rolled into the NDP. The colonel was sitting in the command track where he had been since his helicopter had landed. A short time later, the colonel exited the track and stomped off to his helicopter. Three-seven came to the track and collected his personal items. "That wasn't nothing. I got worse for failing to salute a captain at Ft Knox years ago." Having collected all his stuff, he turned and headed for the colonel's helicopter, which was now at full idle waiting for him to board.

It didn't take long for information to filter out to everyone in the troop. Three-seven had made a comment about the colonel's leadership qualities. The colonel had heard this comment and had court-martialed three-seven as soon as they all got back to F-Troop's NDP. Three-seven was reduced in rank to staff-sergeant and fined thirty days pay. And he had been removed from the unit. This last was the best piece of news I had heard in quite a while. Cherry Boy was crestfallen.

This was my opportunity to get out of the driver's hatch permanently. Before a new platoon sergeant could take over, I asked to transfer to three-eight. The gunner had cycled home, and they were short that position. Every one was agreeable to the arrangement; I was told I could move as soon as new men were assigned. It only took three days. The new platoon sergeant took over along with a new driver. I was shuffled over to three-eight.

Sergeant Groves was the track commander of three-eight. He was a California surfer guy who everyone called Grubby. He knew what he was about but had no interest in busting anyone's chops. Jerry C. was the driver, a big guy with a bigger smile and a wispy moustache. He was also a mechanical genius. With the diesel fuel in the humid climate of Viet Nam, there would be a separation of water out of the fuel. Every couple of days, it was necessary to climb into the turret, twist your body like a pretzel, reach around a retaining mount, and use a built-in pump that had a small ball at its end. Using this you would have to pull out and push in twenty or thirty times to bleed off the water. If this wasn't done, the engine could be fouled. Jerry figured a way to set the pump open just enough to let the water bleed off without having to pump.

On the Sheridans, most of the gunners rode on the bustle rack. Everyone had either an M-16 or and M-60. It came to me that another M-2 on the turret would give me a bit of a leg up. I asked Grove, and he concurred that more was always better in the area of firepower. Jerry and I went to the armorer and asked about getting an additional M-2 mounted on the turret. The sergeant laughed at us. "Everyone wants another 50. I don't have any spares, and if I did I, would have a dozen guys ahead of you."

I pointed to a 50 that had no barrel and sat in the corner of the trailer. It sat on its rear handles up against the sidewall. The sergeant looked at the body of the most fearsome machine gun ever designed. He issued a *harrumph*. "Some jack ass tried to chamber a cold round. It's lodged in there so tight that I can't get it out with my extractor. I'm gonna have to figure something else out."

"If I get the shell out, can we have it?" Jerry asked.

The sergeant guffawed and looked at the machine gun laying, useless, against the wall of the trailer. "Sure, you fix it, and it is yours." The skeptical tone told me that he had every expectation that the useless weapon would be returned in the condition it was in when it left. He was a career weapons man who had all kinds of experience with every type of weapon in the army's inventory.

Jerry looked around and spotting several 50 barrels asked, "Can I have one of those?"

The sergeant looked at the barrels. "Sure, they are all burned out. You can have all of them."

"No thanks, I just need to borrow one." He picked up a barrel, and I picked up the jammed body. I followed Jerry consumed with curiosity about what he thought he could do that a professional armorer could not. We walked over to where a cluster of large rocks sat. Jerry took the body from me and screwed the burned out barrel into it. I watched without a clue as to what he was doing. Once the barrel was properly screwed in place, he picked the weapon up by the rear handles and climbed on top of one of the larger rocks. He lifted the machine gun so the muzzle was pointing straight down at a smaller rock. His arms were out perpendicular to his body. With amazing force, he dropped the muzzle down into the rock. The force drove the barrel back as if it were recoiling. The machine gun's mechanism ejected the spent shell.

As I watched the shell spin out into the weeds, I was forced to utter an "I'll be darned!"

The spent shell successfully ejected, we unscrewed the old barrel and headed back to the armorer's trailer. Jerry handed the spent barrel to the sergeant. "We'll need a good barrel."

The sergeant looked at the body I was holding. "You got that shell out?"

"Yep."

"How?" He looked at the body of the weapon returned to usefulness.

Jerry set about explaining what he had done to get the shell dislodged. When he was finished, the sergeant again shook his head and went to the back of the trailer. He extracted a new barrel. He also picked up a stand and handed it to Jerry. "You'll need this too. If you ask one of the mechanics, they can mount that for you." The sergeant returned to the M-60 he had been working on. The lesson had been learned, and if another 50 came in to his shop, he would know what to do.

We continued our routine of patrols, recon, and road sweeps. I was beginning to enjoy the new environs when my number came up. For the first time, I had to put on the headset and use the mine detector. I had seen the films in training. I had been instructed on the proper technique for probing for a mine, but I had never used the equipment. We started the sweep just south of a tree line that bracketed both sides of the road. I was told to put on the earphone and start sweeping. A funnugie was assigned to probe where I said if I found anything.

There are many things bad about being a sweeper. Not the least is the possibility that there is a sniper out there waiting for you to walk along. When you are sweeping, you don't really have anything by way of protection except your flack vest. It is just you with headphones on and a big dumb grin to ward off the ill will of others. I didn't like the sense of nudity that went with the job. The amount of discarded tin cans, brass, and other miscellaneous metallic debris made the isolation of a tone that actually meant you had located a mine was especially difficult. We had gone along a couple of hundred yards when I got a tone. It was a strong tone, unlike the ones that come from shell casings, clips, empty C-ration cans, and other military debris that littered the roadside. I told the kid who was with me that I had a tone and he should probe where my toe was pointing.

The finnugie dropped to his knees and began stabbing the ground like human sewing machine. I grabbed his hand and pulled it away.

"I said I got a tone dumb ass! Don't stab at the ground like that." I took the bayonet from his hand and began probing like I had seen in the training films. The point came up against some firm resistance. I moved the point back and forth slowly. As I did, I started to define the edge of something round with a rolled edge. I put down the bayonet and began dusting the edge revealing a large land mine. The finnugie, who was kneeling next to me, plopped down. "Whoa, shit!" was all he said.

I motioned, and one of the men on the track clambered down some DET cord and a primer in hand. As we continued our sweeping, they rigged the uncovered mine, and once we were far enough ahead, they detonated it. The finnugie's head snapped around. "What the hell?"

I was intent on the sounds coming up through the headset. I didn't have the proper tolerance for the new guy and his lack of awareness. I wanted to get the sweep done so the feeling of being a target could come to an end. We finally came to the end of the area we were clearing, and the guy that was sweeping the other side of the road and I took off our headsets and climbed back on our tracks. The platoon turned around and headed back to the NDP. We had not gone halfway back when one of the lead ACAVs hit a mine on the side the other guy had swept. I thought as the dust settled that we should have paid closer attention to that side of the road because they always planted mines in twos or threes.

The crew on the ACAV were shaken by the explosion, but none of them were seriously hurt. The ACAV was another story. The outside track was broken. The drive wheel and two of the road wheels were completely demolished, and the side of the track was now badly scared. We set about clearing away the damaged from the track, getting it hooked to another track and setting off back to FSB Marge.

The platoon mechanic took a look at the track and opined that it was a combat loss. In the morning, the ruined carcass would be hauled back to An Loc, and the crew would be sent to Saigon to pick up a new ACAV. Within three days, they were back and part of the platoon's activities again.

When we got back to FSB Marge, the wrecked ACAV still cabled to one of the other ACAVs, conversations opened about how many ACAVs we could afford to total out. We all sat around and threw in our opinions about the cost of all the wrecked vehicles. After having just been at the depot a few months earlier to pick our new Sheridan and seeing all the rows of brand new M-551s there were, I expressed my belief that we could go through one a week and not even breathe hard.

Sheridans, APC, and M-48s were all sitting in the depot, row after row, just waiting for the new owner to come pick them up.

"I'll tell you what: We could lose a track every day and never miss a beat," the wing man of the replaced track said. "They aren't going to run us out of supplies even if they blow up tracks everyday and every night."

The platoon sergeant had been listening to this, and he injected his wisdom to the conversation. "This is a war of attrition, attrition of wills, not of men and materials. If it were about the attrition of men and materials, the war would have been over long ago with us on top." He took a long drag of his cigarette and exhaled slowly. "No! Make no mistake: This is a war of attrition of wills. Are we willing to stick to it until they finally tire of losing all of their people and being economically ruined? That's why when we find those gentlemen in the bush, we have to hold on to them and grind at them until there aren't any of them left. Because, I'm telling you boys something, our will is being challenged, and I'm not sure our people, back home, have that kind of will. Even though most of them won't have to suffer any loss due to this conflict." He stood up from the crouch he had been in and disappeared into the darkness, leaving all of us to think about what he had said.

The next time we swept for mines, we had made about one klick from Marge when a bus full of civilians came upon us. We tried to flag them down to no effect. They waved to us as the driver navigated around our line of tracks. He was about three hundred yards down the road when an explosion erupted and dust rose up in an enveloping cloud around the stricken vehicle. The sweep was abandoned as we went into rescue mode. By the time we got to the ruined bus, most of the passengers were off and away from the bus. Our medic began administering to the injured while some of us boarded the bus to tend to those still on board.

The driver sat in his seat, head back, and blood flowing from numerous wounds. His right leg was torn and bleeding profusely. A man seated just behind the door lied across the seat, several gaping wounds visible to even a perfunctory examination, dead as thanksgiving turkey. A few rows back on the driver's side of the aisle, a young man lay draped over the seat in front of him for all the world like a pair of dress pants hung over the back of a chair. A feel at the carotid artery told me he was dead, but there was not a mark on him that I could see. His white silk shirt was unblemished. He was dead, but there didn't seem to be a why to it. We spent the balance of the day sorting the civilians out and getting the most severely injured evacuated. I, for my part, kept drifting back to the dead man with no visible injuries.

Transport was arranged, and the civilians well enough to continue were taken to An Loc where further civilian transport would take them to their destinations. How extraordinary that the simple act of taking a bus could end up being a fatal activity. And the VC who planted the mine: Was his conscience pricked by the taking of civilian lives, or was that just the cost of doing business? All good questions not meant to be pondered by soldiers in combat. It's alright, ain't no big thing, some would chant after an event.

Chapter Fifteen
SEPARATION

Like a top spinning on a sloping surface, our area of operation moved slowly northwest along the main road. Engineers came and built a new FSB named Eunice. HOW Battery, comprised of six M-109 self-propelled 155 MM howitzers, moved in upon Eunice's completion. The NVA had blown up a bridge on the highway 13, cutting off several towns along the Cambodian border at the boundary of our AO.

The ACAVs were peeled off from the troop and flown by C-130 north to Bo Duc. We were dancing on the sword's edge. Probing for contact with NVA units infiltrating across the Cambodian Border. Everyone knew they were there, in Cambodia, and that they could prepare and recover with impunity. The best we could do was interdict them when they tried to cross. Blowing the bridge had been an attempt to cut the eleventh off from that area giving the NVA a chance to cross and get properly set before having to fight. Our being there by using the Air Force was a direct attempt to thwart that plan.

The Sheridans, because of the modifications to assist in withstanding mines, could not be air lifted into the area. So the troop would have to confront the enemy, if he showed, short on firepower. For our part, we were made perimeter security for Eunice. There were no more patrols. We built bunkers and set about housekeeping.

I didn't think past the luxury of semi-permanent housing and three squares a day.

Bob Hope was coming to Viet Nam to do his Christmas show. Every unit was to get a designated number of tickets. There were to be drawings within each unit, and the drawn names would be allowed to attend one of the shows. We were within days of our show when intelligence came up with a report that we should expect activity in our area, and therefore, there would be no drawings. Groves and I both expressed our doubts about the validity of the intelligence and the belief that some of the headquarters' personnel would be getting our tickets.

The day we were to have gone to the show passed uneventfully; even the communist rats that had taken up quarters amongst us behaved themselves. Word came that the bridge was nearly repaired and we would be rejoining the rest of F-Troop soon. I was not in any hurry to move up to Bo Duc. The rest of the troop had been up there for over a week, and despite expectations, there had been no contact. I suspected they were just waiting for me to go up there before issuing their formal welcome.

On the evening of the 15th, Groves came back to the bunker, and he looked like death warmed over.

Jerry asked, "What's up?"

"We got hit. Big ambush," Groves said in an almost whisper.

"How bad?" I asked. I didn't really want to know.

"Bad. They didn't have numbers, but it was bad."

"Fuck," Jerry muttered. Totally uncommon for Jerry who rarely swore.

It was Jerry's turn at watch, and despite wanting to stay and get more information, he excused himself and went up top, climbing into the cupola.

Groves told us all, "Just because they hit them hard doesn't mean they won't hit us tonight, so be sharp."

I walked out of the bunker and sat on the row of sandbags that protected the entrance to the bunker. There had been a big fire mission a couple of hours earlier. The six howitzers had roared sending their shells out in support of some unit that had made contact. I hadn't imagined that it was our guys they were firing for. Night came like a cloak blanketing everything. I could hear the artillery men talking and prepping to fire. God, I thought, are they attacking them again tonight?

Suddenly the six guns all fired in unison. The boom was like one extended explosion instead of six separate ones. To the east, there was pop, and together six star burst shells blossomed to life, forming a near perfect cross. Only the right arm was slightly below the horizon created by the middle and left arm.

A cheer erupted from the center pit where the howitzers were bunkered. Then there was silence as everyone inside Eunice watched the star bursts sink slowly below the horizon created by the treeline.

There was a sadness and a horror I could not give voice to. My grandpa had once told me that the scariest thing is the thing you don't know. Your mind can create horrors that nothing real can match. So it was that night. I didn't know how many of our guys had been hurt or what the force that F-Troop had fought was. How many of them were there? Would they get reinforcements from Cambodia and come at us even stronger? Would the rest of the troop come south now and leave Bo Duc to the NVA? Or more likely, when would we go north?

The answer came first thing in the morning. We all packed up and moved out on highway 1A toward the rest of the troop. We were joined by D Company, a company of M-48s. The run was a short one. A new FSB had been built on a hilltop that looked cleanly across a grassy plain into Cambodia. The M109s moved into their new firing pit, and we all took up points on the perimeter. Where three-eight parked, we were looking down on an ARVN compound. It didn't take an empath to feel the anger and depression that was bound by the high berm and concertina wire.

Once we had our cots set up and were ready for the night, I went hunting Aggie. Aggie was the TC on two-eight. He was younger than a lot of the other TCs, but this was his second tour. I found him getting back together with his crew. He had been a TC on an ACAV during the separation, and his gunner had run things in his absence. He was from Texas, and he had a warm friendly smile and an ability to tell a story like few others. I found him setting up his cot and giving instructions for guard duty. When he saw me, he raised his hand in acknowledgement and finished his instructions. Organization having been established to his satisfaction, he walked over to me. "What's up?"

"That's what I'd like to know. No one seems to want to talk."

Aggie nodded to a spot on the berm where there was a little larger gap and privacy might be achieved. We sat down, and Aggie looked out toward Cambodia. "I'd really like to take a little ride over there and say howdy to a few of those gentlemen Texas style."

"What happened? How many guys got hurt? I was looking for Mick, and I couldn't find him. The guys on his track wouldn't say anything."

Aggie took a deep breath. "It was ugly. Classic horseshoe ambush. They were really dug in. We'd been down that trail a couple of days earlier and hadn't seen anything. We rolled in, and the first track took an RPG. We herringboned and started firing back. It's been a long time since I've seen that many bullets flying back and forth. They were dug in, and they didn't have any interest in breaking off and running. When it was over, five guys were dead and a gob more wounded. It took us, artillery, and helicopters, but we finally got them to retreat. I haven't seen them so willing to fight. They just really wanted to hurt us. But when they broke it off, they moved off like a stripped-ass ape." There was a long pause before he added, "I guess they'd hurt us enough."

A horrible ache welled up in me. There was a question that needed asking, and I didn't want to ask it, but I had to know, "Who all was killed?

Aggie looked at me, gauging whether he should say anything or let it die. I think the judgment he came to was that he had no right to hold it back. They were my comrades, my brothers as much as anyone's.

"Combs. Skinny little shit took a hell of a marksman to hit him." That made me guffaw unwillingly. Combs was, in point of fact, pencil thin. The smallest flack vest could wrap around him one and a half times. "Neugent, Collins, Roche, and Dawson."

Mick was dead. The one guy I was most sure would make it through all this if justice had anything to do with it. I let my head sag my chin to my chest. "How?" I couldn't ask more. As darkness deepened in the valley below, Aggie detailed the events of the day before. His words landed like blows from a boxing opponent, but I urged him on. The awful story of Mick's death cut the deepest. Then Aggie related what had happened that day.

"We loaded up the dead NVA on the back hatches of our ACAVs and hauled them to the ARVN camp down there." He pointed down to the camp that sat below us. "Tonight they are going to do what all patriots desire; they are going to die for their country again."

"That's our body count. We paid for it," I had uttered the words without thinking the outrage of being denied a fair offset for the loss I felt. Emotions went coursing through me; some I had feared since I set foot in Viet Nam. And the lid I had kept in place on hate came ripping off. As Aggie went on, I could feel a coldness take hold of me. I had been a combatant for four months, but until now, there hadn't been any hate. Now there was hate.

My grandfather was a wise man. He had told me a story about a man he knew who had gone bad. I had observed that it was the circumstances that had made him bad. My grandfather had smiled at me and looked me in eyes and said, "No one can make you bad. You have to invite evil into your heart to go bad. You have to welcome it. If you aren't willing, bad will never find a place to enter." I wasn't thinking of that wisdom; all I wanted was to open up and let evil and hate in. There were bills to pay, and I was now willing to be the debt collector. I hated the NVA; I hated the LIPs. I hated the US Army. In truth there was no boundary I wished to place on it.

Aggie and I sat on the berm looking down. I joined him in silence, thinking of all the people there were to loathe. Aggie tried to change the subject as he entered into a discourse about his tenure at Texas A&M. There was one story about when an Aggie had his boots stolen, and as a matter of honor, many students had attempted to recover the famous stolen boots. I heard the stories, but they could not wedge me apart from the ember I was nursing.

As prescribed, at O 100 the ARVN compound erupted in fire. Streams of tracer rounds flowed in all directions from the center. There were mortar explosions and a few hand grenades. After an interminable period, the firing trailed off and then stopped. The only thing missing in the chaos was fire flying into the camp. Everything was outbound.

"For a bunch a dead guys, they put up a pretty good fight," Aggie offered at last. "Took a hell of a lot of shooting to kill those dead guys."

"Yeah. A hell of a bunch of warriors 'cept, there wasn't any incoming." I didn't wait for Aggie to respond. "I gotta go; my turn on guard will be coming up." I stood, and before I headed off to my track, I told Aggie thanks for telling me what had gone on.

During my turn on guard, I was watching the night sky. With light discipline, it was like a night at the farm. When all the lights were out, you could see millions of stars. It had always been one of the things I loved about the farm. That night I looked at the blackness and ached for my friends. At some point, I noticed red lights moving in formation. They had to be at a good altitude because I could not hear their engines. Suddenly across the now quiet ARVN compound, the darkness erupted into flashes of light and then the distinct thud of high explosive ordinance erupting in perfect cadence. I recalled my first day at F-Troop, and an involuntary shudder went through me. 'Well,' I thought, 'good for us.'

The next day, we went on a recon. Our job was to access the damage of the previous night's B-52 raid that I had observed. It took all morning to get to the site. When we got there, I was properly awed at the destruction that had been wrought. A row of craters fifteen to twenty feet deep abutting each other ran in a straight line through jungle and into open grass land. It had definitely been an encampment. There were partially demolished bunkers, their log roofs twisted upward and away from the bottom of the bunker. There were also untouched bunkers that sat along the fringe of the assault. Anyone who was in one of these would have weathered the attack with no harm.

We spent an hour looking for anything of military value and found little. I found myself wondering what it must be like living through an assault of this magnitude. The sheer terror that must accompany being in the field of attack was unimaginable to me. The final assessment was that the raid had been effective in breaking up the emplacement. The unit that had occupied the emplacement had effectively cleared the site and had left little of value behind. There was no way of developing an accurate body count. There was, in point of fact, no way of even knowing how many bunkers had been destroyed. The examination complete, we loaded up and returned to the FSB.

When we had finished eating and had prepared for the evening, I set out to find Joe, our Chu-hoi. I had to know if he had ever been through a raid like the one we had just inventoried. I found him cutting the hair of one of the guys from second platoon. When he saw me approach, he smiled and asked if I was there for a haircut. I shook my head and said no. "I wanted to ask you if you ever lived through a B-52 raid like the one we just saw."

Joe shook his head. "You get haircut, we talk."

So it was to be extortion. I would have to get a haircut to gain some insight into what the experience was like. I nodded agreement and sat down to wait my turn in the barber's chair. Joe continued to employ the hand shears and comb to the other man's hair. With close precision, he removed small tags of hair, evening out the cut until at last the guy he had been working on looked as if he had just stepped out of a barber shop in downtown New York. He accepted the five dollars offered and then nodded for me to step up.

I allowed him to get into a rhythm with his barbering before I asked, "Did you ever go through one of those bombings?"

"Yes," he said without indicating he would offer any details without specific questioning.

"How many?" I probed. I was irritated that he would make me pull out the details, but I didn't want to get him to a point where he wouldn't talk at all.

"Three time."

"I imagine it was pretty scary." I offered.

"First time. Not know what it was like. Had been in artillery shelling. I thought this same same." He allowed his hand to contract and expand in smooth, even strokes that removed an inch or so of my hair with each contraction. "Not same same. When over, those left came from our bunkers. We all see what was left. I think that numbah ten. My sergeant he say all same. Your bunker hit or not you think too much it do you no good."

He continued to clip away, saying nothing more. After a minute of silence, I asked, "What about the second and third raids?"

Joe paused in his barbering. It was a long enough pause that I started to turn around to see if he was still there. I had turned my head a quarter turn when I felt a hand take my jaw and rotate it back to looking straight ahead. "Numbah two time I coming back with my squad. We small way when raid start, and I got knocked down. Hit head, and I don't know anything afta. I wake up, and all men I was with are dead. Camp place, all gone. Ten, maybe fifteen men live. Rest all kill or hurt." Joe allowed that to sink in before he continued. "That not bad; I not remembah anything. Last time we get no warning planes coming. We all get down in our bunkers. Because ARVN unit on patrol too close. So we get down and raid start. My body feel like some one punch too many time. Head pounding and chest feel much pain. When planes fly off, ARVN come in to area to check damage. No time for get bettah. We have to fight and retreat. We get away, but I think no more raid, no more fight no more listen to people say we winning. I see too much. I say bullshit I quit. I Chu hoi. I here now. Still fighting, but no more airplane raids. This side more bettah."

Joe finished the haircut, and I paid him. I felt like I had gained some insight into the enemy's mind. Not that I cared all that much what those assholes were thinking. I did have them as first on my list of those I loathed and hated sharing the planet with. They were the enemy, and anything I learned was not to develop sympathy for them but to learn how to kill them with greater efficiency.

We continued patrols covering north and west of the new base. In the evenings, an inept NVA mortar crew would lob shells at the camp. They invariably hit short or long of the perimeter, and we would all watch and count off the hits. There were never very many, at most a half dozen, and it was an interesting exclamation point to the day.

We had been intensifying our patrols to the west, working close to the Cambodian border. Third platoon was moving through tall grass while the squadron commander flew overhead monitoring our progress. Suddenly we were ordered to swing on line and turn west. We all knew we were close to Cambodia, and heading west was just going to move us into forbidden territory. The NVA could sit on their side of the border and lob shells at us all they wanted. There was the added danger of moving on line through tall grass. Even sitting on top of our vehicles, we could hardly see each other.

Even a man with no tactical skills could see this was a recipe for disaster. Radio chatter picked up as each track commander picked up on the situation and chimed in his concerns. Each comment was met with an order to stop the extraneous talking and continue our movement west. Then came the order to pivot south. The radio chatter picked up anew. The various comments were met with an order to be quiet as the helicopter swooped in. One of the door gunners let loose with several short bursts from the mounted M-60. With the first burst, fifties on every track jacked a round into the chamber, and the adrenaline roared through everyone. The Huey came to a hover forty meters in front of the line of tracks and then settled to the ground.

When the chopper came to a hover, we could all see that the man behind the door mounted M-60 was the squadron commander. One of the enlisted men jumped out of the side door and ran off to the south. Soon another enlisted man joined him. They converged on a spot, and then in front of our line they picked up the dead carcass of a marsh deer. They carried it over to the chopper where the colonel examined it before having it lifted into the helicopter.

My head started to throb. We had been used as freaking beaters driving game to the hunter. And some hunter, machine-gunning a deer from the air. There was precious little sport in that. On top of the lack of sportsmanship, there was the danger we had been put in driving the deer into an easy shot or, more accurately, shots. I had just found a new reason to hate management.

We began progressing north of Bo Duc in our patrols and sweeps. For several of these actions, we paired up with D Company, three platoons of main battle tanks. When we came into a NDP one night, one of the men from the tank parked next to us snorted something about us being toy tanks. I responded that we might not weigh as much as their vehicle, but our main gun was fifty percent bigger than theirs.

"Yeah, but what does it do bigger other than fart after every round?" came his response. "And not even an excuse me afterward," he added

We had been exchanging barbs about the relative merits of our vehicles when another group of vehicles came into the perimeter. A Sheridan pulled into the gap between three-eight and the M-48. Cox was in the commander's cupola, and when the engine shut down he removed the belt from the M-2. I wasn't really paying attention to what was going on. A thump sounded out, and the ground swelled between my feet. I looked down to see a perfect circle with wisps of smoke rising from it. All of the fear, anger, and hate welled up in me, and I started climbing the front slope of the Sheridan. Cox had to die, and I was more than willing in that moment to administer the sentence.

Three sets of hands grabbed me and held me back. In truth if I had gotten to Cox he could have probably pounded me into rubble, but that thought never crossed my mind as I struggled to get to my objective. It didn't take long for the struggle to extract all my emotions and allow them to ebb. Some one whispered in my ear "Let it go, he isn't worth it." A stronger truth was never to be heard.

That was the last night of joint action with the heavies. In the morning, we would pack up and prepare to move. What had been so important that it had caused the troop to split down making it weaker and more vulnerable was now to be abandoned to the ARVNs. We had bigger fish to fry.

Chapter Sixteen
TAY NIHN

The name went through the unit like water through a fire hydrant. Every muscle tensed, and the minds were sent racing with the desire to recoil. Tay Nihn: the province of death and destruction. At first I thought this was just a rumor. Surely we would not abandon an area we had paid so dearly to defend. But no, it was not a rumor. Two days after the rumor percolated to the surface, we were packing up for a road march.

We retreated down highway 13 to An Loc. Having reached this sanctuary, we assembled in a vacant corner, making it our motor pool for the purpose of maintenance and rest before we resumed the trek to Tay Nihn. We had settled for the evening, and I was sitting on the front slope of three-eight when Sergeant Biese walked past, his head down and a scowl on his face. I had ridden with Biese a couple of times when he had been shorthanded. He was a lot like Groves, good about his job and not interested in throwing his weight around unnecessarily. "Hey Biese what's up?" I called to him.

Biese stopped and looked at me; clearly he was weighing whether or not to tell me his tale of woe. "Aw, someone stole the radio out of my tank. First Sergeant says I find a replacement or I pay for a new one."

"So where you gonna get a replacement?" I asked.

"I haven't got the first notion," came the dejected reply.

"I have an idea, let's go." With that I slid off the tank and fell in next to Biese. We were almost clear of our area when two other guys spotted us. O'Haver was a big guy. He was also older than most of us. He was married with kids. He and the other man fell in with Biese and me.

"Where we going?"

"Biese had his radio stolen so we going to get a replacement," I stated with absolute certainty.

"Where you gonna find a loose radio?" the other guy asked.

"I know just the place!" I advised him. In point of fact, I had noticed a whole row of Jeeps all with radios earlier when I had run over to our troop HQ. The perimeter was protected by concertina wire, and what I hadn't noticed was that all the radios were chained to their mounts on the rear of the Jeep.

"Great, how we gonna get in there, and how are we gonna get a radio off those Jeeps when they're chained?" It was almost a moan when Biese said it. "Forget it; besides, this is the MP compound. Those guys down by the shack are MPs. I don't want to mess with them."

"Well this is where the most radios out in the open are, and we're taking one." I was surprised at how firm my statement sounded. "Come on, we need to borrow some bolt cutters"

By the time we reached F-Troop's HQ, it was starting to get dark. Third platoon's old lieutenant was now the executive office in the HQ. He was sitting at a table with four other guys including the troop mechanic. "What's up fellas?"

"We need to borrow a pair of bolt cutters," I responded.

"What for?" the lieutenant asked.

"We have a chain that needs cutting."

"A chain tied to what?"

"Tied to our missing radio."

"Your missing radio?"

"Well in the technical sense, it's the army's radio, but it really belongs in two-eight, and it got misplaced."

"So you boys are going to un-misplace it?"

"That's the idea."

"The RAM theory. Relocation of Army Materiale. I like it. Do we have a set of bolt cutters?" The lieutenant asked looking at the mechanic.

"Yes, sir."

"Well then we need to lend these guys the tools necessary to accomplish their mission." The sergeant rose and left the table returning in a couple of minutes with the most formidable looking pair of cutters I had ever seen. He handed them to Biese. "I want these back before you apes leave tomorrow.

The cutters in hand, we returned to the MP compound.

"Those boys are gonna be pissed in the morning," O'Haver offered as we crawled up to the concertina wire. A few quick snips, and we were inside the motor pool. There were three MPs clustered around the light of the shack that sat at the entrance to their motor pool. We approached the nearest Jeep, and Biese looked at the radio. He shook his head no. We moved to the next Jeep in line. After a cursory exam, Biese gave the okay sign. I took the cutters and applied them to the chain. The jaws of the cutter bit into the link and was through with amazing ease. Disconnected, the chain started to slide off the top of the radio, making a metallic clanking sound. A moment of terror seized me. I was sure that HQ in Saigon had heard the chains clank. Biese grabbed the end and lifted it free. Once it was away from the radio, O'Haver took the handles on either side of the radio and gave a huge jerked upward.

The whole back of the jeep jumped as if it had hit a pothole at speed. The loud clank issued from the area we were working. We all fell down and waited for the MPs to come inspect the area. To our relief, the three MPs didn't even throw a flashlight beam in our direction. After an eternity of waiting, we set back to work.

Biese whispered, "You need to undo the mounts on the mounting plate."

"Right," O'Haver whispered back. Four easy turns and the clasps that help the radio to the plate were undone. O'Haver raised the radio clear of the mount. Biese reached out and undid the antenna wire and the power cord. Clear of any further connection to its previous home, we took the radio and exited through the gap we had made in the concertina wire. The other guy returned the bolt cutters and advised the lieutenant and the sergeant that it would probably be best if they denied all knowledge of the evenings activities. O'Haver, Biese, and I made our way back to our area.

Biese's smile was bright enough that it lit the way back to our motor pool. Within ten minutes, the radio was locked in place and a com check was made. It worked perfectly. Biese offered a heartfelt thank you, to which we nodded acceptance and went back to our individual tracks. "Blackhorse" was all that was said in response to the evening's events.

The next morning, we assembled for the next leg of the road march. As we were starting through the main gate, we were stopped by six MPs. They clambered up on the first ACAV in line and began messing with the radio. After finding the serial number and deciding it didn't match, they moved to the next ACAV. They had just finished and were about to climb up on the next vehicle when the first sergeant stopped them. "What do you think you're doin'?"

The ranking MP, a staff sergeant, looked at Top and smiled. "Someone stole a radio out of our motor pool last night. We're checking every vehicle that goes outta here until we find it."

"Uh huh." Top looked down the line of vehicles. "Gentlemen, I have a schedule to keep. It ain't my schedule; it belongs to my colonel. We have a long road march, and you aren't helping me any. Now if you like, I could get my colonel on the horn and you can explain to him why his unit isn't going to get to their NDP before dark and why they won't have a secure perimeter tonight. Or you can let us get on our way. Don't make any difference to me. Although I haven't watched a good ass chewin' for a while, and I'm always up for one long as it ain't on me."

The staff sergeant looked hard at Top and decided there was no bluff in him. And the last thing he wanted was to be toe-to-toe with a pissed-off colonel. "Get your people moving." He motioned to his cohorts to get down. One protested but was quickly dismissed out of hand. Under the watchful eyes of our first sergeant, F-Troop marched peacefully out the main gate. As two-eight passed the formidable first sergeant, he smiled and gave a quick salute.

Places have different feels. This is never more true than in a combat zone. If I had been nervous before, it was nothing in comparison to my feelings now. On the march, we passed the bone yard: a perimeter of burned out, wrecked hulks of military vehicles sitting in a circle defending an empty patch of ground. Here a valiant defense had been mustered against an enemy who cared not what their losses were as long as their opponent suffered total loss. If he lost 5,000 men to kill 100, that was victory as long as the 100 was everyone. That mindset puts a new measure of terror into your calculations. This was Tay Nihn.

There was a different feel to this place. Most notable was the absence of LIPs. Where we had been patrolling before, there were always locals willing to conduct commerce with us. Here there was none of those activities going on. This added to the sense of dread. The locals had near perfect radar, and where they weren't the enemy probably was.

On top of the dread created by our new environs, Groves cycled home. A new sergeant came in and took over three-eight. At first blush, he seemed ok. He didn't come in looking to bust anybody's chops, and he seemed to acknowledge his newness to the tank and the troop. The three of us were willing to give him the benefit of a doubt.

The third day in Thay Nihn, we pulled into a Green Beret base camp. In addition to the Green Beret combat team, there was a battery of Long Toms, 175 MM artillery pieces. We took positions on the perimeter while Captain Bailey worked out assignments with the Green Beanies. Word quickly spread through the troop that there were Vietnamese hookers in the encampment and they were open for business.

After dinner and our cots were all set up in preparation for the night, a line formed outside the comm trailer that was serving as a bordello on wheels. Rumor had it that the girls were very attractive and especially capable. For those of us not inclined to such activities for pay, there was money to be made sitting guard for guys who were in line. Cherry Boy came to me and offered twenty dollars to sit his guard post. I was more than willing since I really didn't expect that Cherry Boy would actually go through with it.

Shortly before his turn at guard duty came to an end, Cherry Boy came to thank me for filling in for him and to advise me that he had been with one of the girls. He told me how pretty she was and what an earth rocking experience it had been. I told him I was glad to have been of help in his quest for manhood. What I really wanted to tell him was that being here in a combat roll in a foreign country made him way more a man that any sexual experience ever could. In the morning, we packed up and headed out toward our AO.

One of the first actions when we arrived in the province was to sweep the roads. We came to a road that was going to have to be swept by us. I knew who was going to be put on the detector, and the prospect made me weak in my knees. There was a reputation that this province had that challenged thoughts of bravery. The thing was when we got to the starting point we saw three mounds. Two were on the right side, and one was on the left. They were spaced about fifty feet apart, and they defined points of a large triangle. We dismounted and began moving dust off the first mound. Beneath the mound of dirt sat an anti-tank mine. While some of the men rigged the first mine with C-4 and DET cord, several of us moved to the second mound. Beneath the dirt lay a second mine identical to the first.

We cleared all three mines by detonating them in place. Once those mines had been cleared leaving large craters, we proceeded to sweep the rest of the road. I heard tones, but all we came up with was miscellaneous refuse from years of traffic up and down the road. When we were finished, our new TC looked at the stretch of road and laughed. "Looks like they're hurting for trained personnel. Must have had to raid staff from the three twenty third mess-kit repair company to mine the roads." It was a knock on the reputation that this area had for being home to fiercest and best-trained troops the North could muster.

We had been working the area, becoming acquainted with our new environment, when we received what I considered the heaviest blow of all. Captain Bailey was being cycled out. He was being given a new job at headquarters. For officers there are multiple legs to building a successful career. The captain had clearly succeeded in building a strong leg of combat command. Now it was time to let another officer take the reins and show what he was capable of. In the meantime, Captain Bailey would show that he had an understanding of what it took to run a larger unit.

I liked Captain Bailey. He had been a no nonsense commander with a clear perspective of his duty and how to conduct it. He was about the mission not stepping on people's toes. He also displayed an affection for the men under his command. He had our back, and as a result we had his. A new six came on board, and without fanfare Captain Bailey departed.

It did not take long for the new six to show that he had no comprehension of how to run a combat unit. Where in the past one platoon was always kept back to guard the headquarters tracks, we now were all dispatched to cover various duties assign to the troop. On top of leaving the headquarters tracks unprotected, we had the platoons out running around in circles.

We had been a week under the new CO when we were probed at night. It happened to be my turn on guard. As I sat in the cupola, I didn't see any movement along the perimeter, but the ping of a bullet ricocheting off the front of the tank definitely caught my attention. I began firing bursts into the tree line to my left. The whole perimeter was returning fire. I couldn't see anything coming in, only tracers arching outward from every track on the perimeter.

Our new CO called in air support. A pair of Cobras began lacing the edge of the tree line with their miniguns. Solid lines of red waved back and forth like a garden hose saturating the tree line with 7.62 ammunition.

The miniguns stopped, and the Cobras made a wide turn. I heard a *swoosh* and an explosion that caused my right ear to start ringing. Dirt and foliage showered down on me as I realized that the Cobras had just let loose with their rockets. I keyed my mike and started screaming for the headquarters track to call off the Cobras.

Several other radios crackled out the same order. It didn't take long for the Cobras to get the message. They broke off before they could fire another salvo. Within a minute of the helicopter assault on our position, we were ordered to cease fire. The machine guns fell silent, and everyone took a minute to take inventory and thank the maker for the quick action of someone on the command track. Three-eight climbed up to the cupola to ask me what had happened. I pointed to a smoldering crater where a rocket had struck the ground just a few feet in front of the front deck of the Sheridan. A low whistle was all I got in response.

The next morning, word quickly spread that it was out new CO who had given the Cobras the coordinates for firing their rockets. Someone opined that he would eventually get someone killed and it was more likely sooner rather than later. In the afternoon, the first sergeant left on a helicopter. Where he went or why was left blank. In honesty I didn't even realize he was gone. We went out on an ambush patrol, and for once I was glad to be away from the main encampment. The further away from six the better.

The next morning, we came in from patrol in time to see a Huey land in an open area to the north of our perimeter. The first sergeant, Captain Bailey, and the squadron CO dismounted and headed directly to the command track. They weren't in there long before six, and the colonel marched back to the Huey and boarded. The colonel looked flushed and was clearly not in a good mood. Six looked like a child who had just gotten back a term paper marked with giant red F. I didn't care; it was apparent that Captain Bailey was back as six, and that was good news in any language.

Word spread quickly that, against all military protocol, first sergeant had gone directly to the colonel of the regiment to get the new six relieved. Looking back over the years since, I saw First Sergeant Johnson as the prototypical non-com—a man dedicated to the mission and to his men. Whatever the actual action, we had Captain Bailey back, and there was a sense of relief that surged through me.

Our sweeps continued but with a feeling of organization and purpose that had been missing the previous three weeks. We moved into one area where activity had been spotted. Coming to the area, we all dismounted and began searching the ground. There were clear indications of activity, trampled foliage, carelessly discarded paraphernalia, and mats used to cover spider holes. We were looking for anything that might provide intel when a short burst of AK-47 fire cracked through the surrounding jungle. The short burst of half a dozen rounds stopped and within seconds a longer burst of M-16 fire responded. There was a scream for a medic. Everyone fell back into a defensive perimeter. One of the men had taken three rounds to the abdomen, and he lay prostrate. We immediately began flattening the brush with the tracks. In short order, a landing zone had been carved out of the thick jungle.

In a blink, we had created a perfect landing area for the dust off. The landing area had been completed just as the dust off came into view. Instead of making an approach to the newly created landing pad, the helicopter circled on station over us. Captain Bailey called for the helicopter to come in, but the pilot responded that he believed we were still under fire. The captain's response was direct and to the point. Either the dust off would come down of its own volition, or he had some thirty machines guns that he would turn loose on the reluctant pilot and he could come down the hard way.

While the captain was engaged in negotiations with the pilot of the dust off, the wounded man lay writhing in pain and yelling to the captain about peace and making the V sign with his right hand. Doc and two others struggled to calm him and keep his writhing in check. As the captain made his standing with the pilot clear, he made the peace sign to his wounded man. Someone suggested another dose of morphine to aid the wounded man. Doc shook that off with a statement about overdosing the wounded man. His cries of pain and disbelief mixed with the frustration with the evac pilot's refusal to come in to pick him up made a cauldron of boiling emotions.

Fortunately the captain's point had been made, and soon the Huey was entering a hover in the center of the just crafted helo pad. The medic and another man got the stretcher into the gaping cargo door with the large red cross, and soon the men in the dust off had the cot strapped in and the helicopter made a high performance takeoff. All eyes on the ground watched the Huey as it nosed down and began to make speed and altitude. The sight of a man laying wounded and struggling to hold to life had been very sobering. But the knowledge that our CO had our back regardless of the situation made the road a little smoother.

Several days after the incident with one of our guys getting shot, we went on a recon with E-Troop. We were driving up two sides of a peninsula of jungle. The hope was that we would catch some of the enemy in the trees between us. I was standing in the loaders hatch behind the M-2. We had been told there was a strong probability of contact. As always I believed the reports. In my mind, everything we did had a strong probability of enemy contact. Being told so by some jerk in headquarters neither reinforced nor diminished that perception.

We had been moving at a crawl trying to keep parallel with each other. I heard the shot, but I couldn't tell where it had come from. I didn't need an invitation. My thumbs came down on the butterfly with all the force I could muster. The trigger for the M-2 was a metal plate that sat between the two handles. It had the shape of a butterfly with its wings spread to catch the sun. Unfortunately, my thumbs had too much force behind them, and I broke the butterfly. After the first round belched out the muzzle, I had no way of firing.

I looked around for an alternative means of firing the M-2. There were none that I could see. Rubin had the M-60 and was washing an area off of the left rear quadrant. I found an M-16 and picked it up. I charged a round and was about to fire off a burst when we were ordered to cease fire. Everything went quiet except for chatter over the radios. It took a while, but finally the information filtered down. Apparently a NVA soldier sitting in the middle to the two advancing columns had popped a round off at each side. This had evoked an immediate response from F and E troops. We were essentially shooting at each other. By some miracle, no one on either side of the tree row was hurt in the ensuing melee.

Territory was becoming more familiar as we swept our areas with regularity. There was a lot more jungle busting in this province. There were fewer plantations, and cleared ground was less common. We were plowing through some rather thick jungle with bamboo groves. One of the tracks was moving under a large stalk of bamboo that leaned a forty-five degree angle across our route. From where I sat on three-eight, the stalk looked pretty solid, but when the T's helmet hit it, a split opened up and a huge black mass fell down into the turret through the commander's hatch.

Within seconds four bodies piled out of the tank, and clothes started flying in every direction. Men were swatting at their torsos, arms, and legs. Three-eight realized what was happening, and he grabbed the fire extinguisher. Soon the four naked men were surrounded by men firing their extinguishers at them. It took several minutes, but at last the ravaging ants were quelled. The four hapless men stood there completely indifferent to their nakedness. Doc came up to each man and applied salve to the swelling bits.

The remainder of the patrol was cut short, and we returned to the troop compound three-six reported contact with an army of enemy ants. Six took the report with a smile and advice that next time it might be best to circumvent bamboo groves where hostiles might be laying in ambush.

It was practice to move the troop encampment regularly lest the enemy set up a mortar crew and launch, at a minimum a shelling at the worst affect a coordinated attack between mortars and sappers or ground units. We had just established a new perimeter in a grass-covered, open area. A lone tree stood at one end of an elongated oval that was pinched near its middle. The second day after we established this new compound, we went out on a sweep that took us across a well-traveled trail. Note was made of its location and the fact that it was well trod. Three-six determined we would come back that night and set up an ambush.

We returned to the troop compound and waited while three-six laid out his plans for the night. It was beginning to grow dark, and the platoons that were staying in had men going out to plant trip flares. Three-seven sat next to three-eight, but unlike three-eight, there was no one on the tank. I was writing on my boonie hat the places I had been. It was now six months since I had arrived in country, and I wanted to remember everywhere I had been.

I heard the *whoosh*, and then I felt the concussion from the explosion. Fire erupted from the commander's and the loader's hatches. They looked for all the world like a volcano belching heat and fire. I looked across the burning vehicle and saw a puff of smoke just inside the tree line to the south. I grabbed the M-60 and began firing in that direction. Soon the whole perimeter was in response. The guys who had been out planting trip flares were caught in the midst of the conflagration. Some were running back toward the perimeter while others fell to the ground looking for sanctuary of lower altitude.

Three-eight was quickly in the commander's cupola spraying the area ninety degrees from where the RPG had come. Instead of short bursts, he just held the butterfly down sending a stream of rounds out against the unseen enemy. He burned through the first two boxes and reloaded. I had stopped with my second box, figuring that nothing more was incoming and our assailant was either dead or safely escaping into the thick jungle. Halfway through the third box, the M-2s barrel burned out, and tracer rounds began spiraling in undirected corkscrew trajectories. Some flew high; others augured into the ground a few feet in front of the tank.

Despite the clearly ineffectual results from shooting through a burned-out barrel, he persisted. At last we were ordered to cease fire. I grabbed three-eight's arm and yelled in his ear that we suppose to cease fire. He continued firing until the last round in the box was arching over the tree line. When all of the guns went silent, the men who had been out laying the perimeter came in. Several trip flares sat burning out in the tree line the victims of our fusillade. By some miracle they, along with the stricken Sheridan, seemed to be the only victims of the assault.

Third platoon packed up and headed out to the trail we had discovered. As we rolled back on the path we had created heading back to our compound, I realized we had created a perfect invitation for anyone who had been trekking down the trail. Seeing where we had crossed, they had to simply follow our tracks and *boom*, literally. I looked around the bustle rack and observed that we were down on ammunition and our fifty caliber ammunition would do little good tonight. I prayed, silently, that there would be no encounter that night. My prayer was answered.

Chapter Seventeen
ENCOUNTERS

Changes came in dribs and drabs. Rubin changed over to an ACAV. He didn't say so, but I surmised that he was not enthralled with three-eight. We got a new kid fresh in from home. He was Polish and from Chicago. He never seemed to have a lot to say, and I was fine with that. We continued to patrol and encounter little in the way of enemy activity.

Even in the midst of terror, there can be boredom. It is a paradox that I cannot explain. The repeated activity with long periods of no contact lull the sense of terror from a roaring tiger to a hissing house cat. We covered the same areas and failed to encounter the enemy. We did encounter local wildlife. We were clearing a back trail one day when the lead track reported there were turkeys in the road. Six responded "Turkeys?"

"Affirm sir, about half a dozen."

"Turkeys?" came the second response.

"Well maybe not turkeys, but they are big ass birds."

"Cookable?"

"They look tasty to me sir."

"Well it's about lunch time; let's light 'em up."

The lead track opened fire on the flock. Birds took off in all directions, and machine gun fire poured forth. We managed to bring down three of the turkeys. They turned out to be peacocks. No matter, a fire was quickly prepared, and bamboo spits had the birds cooking over the roaring fire. Jerry, it turned out, was in addition to being a mechanical genius a gastronomical wonder with a spice rack tucked in his driver's compartment. We all took part in the feast remarking how much like chicken peacock tasted. Our meal completed, we headed back down the road all wearing fresh plumage and howling like Indians.

The destruction of three-seven notwithstanding the lack of real contact began to make us believe that perhaps Thay Nihn's reputation was undeserved. Contact was sporadic and usually was made with just one or two enemy soldiers. Bo Duc had been much more lethal. That impression changed on March 5th. We had set up a new encampment and set our perimeter. I was on guard duty at ten. I hadn't been in the cupola too long when I heard an explosion. A couple of tracer rounds flew over my head, and I dropped my thumbs on the butterfly. There was a dark spot that I hadn't noticed when we were setting trip flares, but it hadn't moved the whole time I had been on guard. Now that shooting was opened up, I started peppering the dark spot. Tracers disappeared into the lump, but it didn't move or show any signs of animation. Another explosion and the waning fire picked back up in intensity.

At last cease fire was called. All the guns on the perimeter grew quiet, and the distinct *thump, thump, thump* of a Huey's rotor blades filled the air. A small clearing off to the left of three-eight filled up with the body of the Huey with a big red cross on its side. I saw a man limping toward the hovering aircraft. As he walked, I looked around the bustle rack and found a full can of 50 caliber. By the time I had the new box of ammunition in place, the Huey was lifting vertically and then moving off toward An Loc.

It wasn't until the next morning that I got the word that there had been one fatality the night before. The quintessential soldier had been killed. First Sergeant Johnson was, to me, the best soldier I had encountered. He had a no nonsense approach to his job that made it clear there would be no tolerance for slacking off. But there was never a need to scuff the shine on a man's boots just because you out ranked him. If Captain Bailey was the mind that directed the troop, First Sergeant Johnson was the soul that gave reason to the madness. He was the one who provided the how to the captain's what.

He had exceeded all bounds of military protocol when he had gone to Second Squadron Headquarters and gotten Captain Bailey reinstated as troop CO. It would have much easier to have pulled back and let things unfold. He was not that kind of man. I had a horrible sense of loss.

Our mission changed. A bunch of Rome plows were deployed to Thay Nihn, and our duty was to stand guard over them as they set about the task of moving the jungle back a hundred meters from the road. The Rome plows were large caterpillars with flat blades that were a bit wider than the tractor and half as high. A large steel cage encompassed the driver's position affording protection from falling foliage. The reason we called them Rome plows was because they were built in Rome, Georgia.

I had it though with the job of sweeping for mines; there was nothing more scary and dangerous, but what I now observed was even scarier. These men would plow down trees and clear underbrush. There only protection was our tanks and ACAVs that would stand back from them a good fifty meters or so. Each man had some type of personal weapon. The scariest of which was a sawed off pump action shotgun that one operator kept strapped next to him on his seat.

Torn down trees and shrubs were piled then sprayed with whatever accelerant was handy, usually diesel fuel. The great pile would then be lit to burn down to ash. We had cleared about two klicks when the plows turned from widening the road to creating a huge open area with a dirt wall. Inside the wide perimeter, a set of berms were plowed up. Huge ditches were cut. After three days of effort, Fort Defiance was complete.

Into one of the interior perimeters a pair of eight-inch howitzers were placed. The six M-155s of HOW battery were also brought in. One of the large trenches that had been dug became the ammunition dump for the artillery. Two dusters, twin forty millimeter canons mounted on a tank frame in an open turret, were brought in to bolster the perimeter. As I observed this build up, I found myself wondering what kind of contact were we anticipating?

One of the things that got all of us through was the idea of R&R. Sometime in the middle of your one-year tour, you could take a break from the carnage and go to one of the approved locations for the brief vacation. Hong Kong, Bangkok, Sidney, and for married men, Honolulu were the approved venues. Landholm was a married guy who had been in country three months less than me. He took R&R to Hawaii. When he came back, he was a lot more sober in his demeanor. He told all of us that getting back on the plane to come back was the hardest thing he'd ever done.

I had seven months in the country, and I had been putting off scheduling my R&R. Now my reluctance was set in stone. I wasn't sure that if I left this place I could ever make myself come back. "No," I told myself, "I would ride this train to the end of the line." When I left, it would be permanently. There would be no coming back for me.

Once Fort Defiance was complete, the Rome plows returned to their original assignment. Like leaf cutter ants set into endless motion to accomplish the task at hand, the plows moved about leaving stubble and bare earth wherever their blades touched. We were at this job of monitoring the plows for over a week when three-eight shot craps.

We were sent back to Fort Defiance to have a new engine and transmission put in. Having watched its construction and seeing the artillery pieces put into place, I had a sense of comfort knowing we were going inside such a forbidding fortification. The prospects were made even better when we were pulled into a slot under the berm that housed the eight-inch howitzers. No guard duty for at least a couple of days. I was approaching my eight-month mark. Two-thirds of the way through my year. The first evening, a fire mission was called for the eight inchers. As we watched, we could see the huge shells arching out through the darkening sky. I actually had pity for whatever was going be on the receiving end of those loads of explosives and shrapnel.

The second night we were lying on our cots. Ski had his radio tuned to AFVN, and Michael Jackson was inviting a girl to lay down, girl. Three-eight smirked something about a kid not even in puberty singing lyrics like that when there was a loud pop. It was coming from the other side of the compound. When the second explosion erupted the shout of "IN COMING" came up from everywhere. I rolled out of my cot and grabbed my flack jacket and helmet. I scrambled up the side of three-eight and grabbed the M-60 by the handle over the receiver. I grabbed three cans of ammunition and yelled to Ski to grab as many cans as he could carry. As I headed down the rear of three-seven, I yelled over my shoulder for Ski to grab his M-16 and a bandoleer of M-16 ammunition too.

An explosion on the road behind the exterior berm went off, sending dirt skyward and sparks flashing in every direction. Another shell exploded five yards closer. I slid off of the rear end of three-eight and half stumbled across the opening. As I was halfway across the road, I dropped two of the cans of ammunition. I made it to the berm and threw the M-60 up on top of the dirt. I was lying next to an M-88. As I lay there for a second catching my breath, I started running an argument in my head. Should I go back and get the two cans I had dropped or should I set up fire off my one can and then go get the two I had dropped?

If there is any logic when one is in that kind of situation, my logic said the two cans were in the middle of the road, and a mortar round could destroy them both. I turned and headed back to where I had dropped the two cans. A mortar shell that went off a few yards away. I could just make out the two cans lying on their sides. I grabbed them and ran back to where I had set up the M-60. When I got there, the M-60 was lying on its side down at the base of the berm. I could not figure that out; I was sure of where I had left the weapon.

I got the machine gun back up on its bipod and popped open a can of 7.62 ammunition. I got the belt in place and closed the lid. I looked out to the tree line. The thought hit me that Ski was nowhere in sight. I wondered for just moment whether he had been hit. It was either that or he had lost me in the confusion and had ended up with Jerry and three-eight at another point on the perimeter. The M-88 was spraying the tree line in a v with the M-2 mount and its apex. One of the legs of the v intersected with a v laid down by the ACAV that was on the perimeter to my left. I decided to fill the point where these two fields of fire intersect. I squeezed off short bursts. I had fired off a half a dozen bursts when a huge thump filled the air and made my left ear ring. My left arm was up on my helmet, and I was on my right side.

I rolled back and began squeezing off rounds. I reached the end of the first can and went to reach for the second can. It was then that I realized I couldn't feel my left arm from the elbow down. I reached with my right hand, and I could feel blood. I looked, but it was too dark to see how badly damaged the arm was. I realized too that my left side was throbbing like I had been kicked. What bothered me most was that I had no idea how badly injured my arm was. I couldn't tell how much I was bleeding either. I could tell that my sleeve was wet, but that was it.

I had to get somewhere where I could see what was wrong. I ran back across the road and made my way along the berm to the first aid area. Under the flap that covered the entrance, there were medics aiding the wounded. I could see a sergeant first class with a badly mangled arm. I remembered why I was here, and I looked at my arm. To my embarrassment, I could see that my arm was all there. Everything from the elbow down was as dead as roadkill. Blood seemed to be coming from all over the forearm. What I couldn't understand was why I couldn't feel anything. I looked at my hip and side and saw blots of blood from below the joint to up above it.

A medic looked at me. "Where you hit?"

"My arm, my back, and my hip. But I can't feel anything in my arm."

He looked at the arm. Then he swiped it with a gauze. "You'll live." He looked around and then turned back to me. "Come over here, you can help out with some of the more severe wounds." The comment wounded me. I wasn't a malingerer. The fact that I couldn't feel my arm was testament to my being wounded.

He had me sit next to a man on a blanket on the ground. He was breathing heavily, and his shirt was wet with blood. The medic took the gauze in his hand and wiped the bare chest. Bubbles from several points on the chest began to form. "Sucking chest wound." He went and retrieved a poncho. Together we managed to wrap the chest tightly with the poncho. The man's struggled breathing eased a bit. "Hold this tight around his chest. Don't let it loosen, or he's going to be in trouble again. I'll be back."

He turned and left. More wounded began appearing in the first aid trench. I watched as men in various states of injury filtered in. Some looked less injured than me, but most appeared much more severely hurt than me. One came in on a stretcher, his arms flopped over the edges. The doctor took a look at the man. He put his stethoscope to the man's chest. He moved the bell to several locations about the chest. He turned to retrieve a blanket and placed it over the man's face. "Put him over there," he said pointing to a corner that was unoccupied.

Now guilt swept over me. Here I was amidst all this devastation with pinprick wounds while men were coming in severely wounded and some in fact dead. The sergeant first class suddenly realized that his arm was ruined, and he began to react. "Don't you bastards take my arm. Don't let 'em take my arm. Doc, I'm beggin' ya, don't let 'em have my arm." The sergeant didn't seem to be in pain, just the anguish of possibly losing an appendage. I watched all this while trying to keep the poncho tight around my charge's chest.

The study thump of incoming shells abated and then stopped. I could hear the study belch of miniguns off somewhere in the distance, but within the perimeter all grew quiet except for the sounds on men yelling and talking. I heard the familiar *whoop, whoop, whoop* of a Huey's rotor. Stretchers were picked up and carried out of the trench. The man I had been assisting was one of the first loaded into the dust off.

As the trench cleared of injured patients, the doctor turned to me. "Where are you hit?"

"My hip, back, and arm. I can't feel anything here." I swept my right hand down from my elbow to my fingertips.

The doctor twisted my left arm upward to look at the elbow. "Ulnar nerve probably." This guy needs to catch the last evac if there's room. If not, he can go in in the morning." He looked at me. "You'll be alright either way." He then smiled and turned to attend to others who were filtering in.

The last evac copter had room, and before I knew it, I was being whisked away from Fort Defiance. I watched through the open hatch as the dark landscape slipped below the skids. Some one yelled into my ear. "Bad?"

I looked at him and replied, "Bad enough." I was enwrapped with the view of the world below me. Somehow the war seemed a distant thing as we glided along above it. This was how I was supposed to have fought the war. Not down there in the muck and mire and drudgery of combat but over the land flaying in to rescue the wounded. 'Well,' I thought, 'it is what it is and that is all gone now.'

We landed at An Loc and were hustled in to the hospital. I was put on a gurney, and my shirt and boots were taken. A sheet was thrown over me, and I was told to relax. After a minute, a nurse came over. She took my pulse and listened to my chest. She made a cursory examination of my wounds, and satisfied that none were life threatening, she announced, "We have a dirty one here." She then turned to the next gurney up the hallway.

I felt like yelling my apologies for not having showered before being evacuated. A corpsman came to me after half an hour. "We're gonna get x-rays, then we'll see what we can do about those wounds of yours. You need anything for pain?"

"Nah. Like I told the doc at Fort Defiance, I can't feel a thing from my elbow to my finger tips. My hip feels like I was kicked by a mule, but it's ok."

"BS. You don't need to be in pain. I'll get you a shot. "

Before I knew it, I was in La La land. All worries disappeared behind a veil of medication. I don't remember the x-ray. I don't remember being told they were going to do any procedures on me. I do remember becoming aware of someone poking me in the back. I asked what was going on, and a man advised me they were probing for a piece of shrapnel. I told them fine, but don't throw it away. I wanted it for a souvenir. "No problem," came the response from behind me.

The next day, I was evacuated to Than Son Nhut Hospital. I was placed in a ward with men suffering a variety of injuries. One thing I discovered as I was cared for was that I had no tolerance for morphine. I became abusive and nasty under the influence. Someone from regimental headquarters came in and placed a placard at the foot of my bed that had my name, rank, and unit on it. He had mistakenly indicated I was in E-Troop. Later, after the fact, I was told that I had been unspeakably nasty to the man when he came in the next day to see how I was.

One man who was put in the bed next to mine was a burn victim. He had been in a helicopter crash and had suffered burns on his shoulders his feet and hands. Each day they would take him to have the dead tissue soaked and removed. He said when he was first immersed in the bath it felt good, but shortly after the pain would come in waves. I asked him if there was anything I could do, and he said talk. I asked about what, and he said it didn't matter as long as it would help distract him.

I threw myself into the project with full commitment. I told him stories of my life growing up, my grandparents, my fiancé, my sisters, and mom. When I had exhausted the experiences of my short life, I began to weave fantasy with fact, and I also wove in the stories of others I had encountered in the army making them mine. I have no idea whether my prattling did him any good, though he never asked me to stop.

Aggie came in to see me on my third day. He had seen Jerry and three-eight. They had been wounded in the same attack. Apparently they had set up a firing position on the other side of the ACAV that had been to my left. Jerry had shrapnel all over his back. Three-eight had a few minor scrapes. Aggie didn't know whether Jerry would go back out to the field right away or what would happen. I was pleased to hear that he was alright. After half an hour, Aggie made his excuses and left.

After six days in hospital, the doctor came in and looked me over. "You are healing up pretty well. I think we'll send you up to Cam Ranh Bay for about five days. Get a little physical therapy for that hand and elbow." He turned to leave, and then he stopped as if he'd just remembered he needed a gallon of milk in addition to all the other groceries he'd put in his basket. "Your hand still numb?"

"Yes, sir, except for my thumb. I can feel the tip of it like it's asleep you know?"

He took my hand and began poking each digit and asking if I felt the poke. My thumb I could feel at the very end; the rest of the arm and hand were non-existent where feeling was concerned. I had given up even trying to use the hand because without feeling I was constantly dropping or spilling things.

After he had finished poking me all up and down my arm and hand, he looked at the chart. "No, I guess Cam Ranh Bay isn't going to get that fixed. I'm sending you to Japan for further treatment. It's impossible to tell where the damage is here in the country. You'll need a specialist who can properly diagnose and treat that wound."

The statement hit me like a load of gravel being dumped all at once on me. "What are you saying? I'm never gonna feel anything with this hand?" I held up my left hand for emphasis.

"No, I'm not saying that. I am saying we can't treat it here. You need to be in a facility that can provide treatment for that wound, and that is not here." He started to walk away and stopped. "You'll leave tomorrow."

As I was digesting this piece of news, my bunkmate came back from his bath. As he lied there and the pain began to radiate from his wounds, I launched into a story and kept rattling on for several hours. I don't know if any of that did him any good, but it kept me distracted. I was going home! That was the good news, but I was going home having failed in so many ways. I hadn't protected my comrades. I hadn't even begun to avenge my lost brothers. I was a bust like I had been all my life.

The next day, we were sent by military transport to a hospital in Japan. Better food, better beds, and more attentive hospital personnel. I didn't feel much in the mood to celebrate anything. I had several x-rays taken, and a physical therapist began giving me exercises for my hand and arm. I asked him what the point was, and he told me that even though I couldn't feel the hand and forearm, they needed to be moving lest they freeze up.

I got a message from my mom that a kid from Arlington Heights was in my hospital with a spinal wound and that I should go see him. It wasn't something I wanted to do, but I gave in and agreed to see him. The next day, I went about the task of tracking the guy down. I found him in a bed, a halo attached to his skull. I introduced myself, and we began to reminisce about Arlington Heights and our families. He had gone to Arlington High so in athletic events he was the enemy. We had both been to the Cellar, a dance place where local bands played on Friday and Saturday nights. When I said something about being a poor dancer, he smiled and said, "I guess I won't be worrying about that anymore."

I feel like an idiot. After some more conversation about where we had been in Viet Nam and how we had been hurt, I made my excuses and left. I felt like a total jerk. Here I was pouting about a numb arm, and there he was numb from the shoulders down. Selfish, that was all I could think of myself.

The next day we went to the PX. I found a telephoto lens for my camera and bought it leaving me nearly broke. One of the guys I was with bought a gold ID bracelet for his girlfriend. The old man who ran the jewelry part asked if he wanted he name engraved. He said yes and wrote out her first name. The old man bowed and said it would be $0.25 a letter and it would take a couple of hours to do. After we left, one guy opined that the buyer had been screwed on the engraving. $0.25 a letter was way too high a price to pay. The kid who'd bought the bracelet said he was okay with the price and we were to drop it.

When the hour expired, we returned to find the old gentlemen hunched over his workbench, a small chisel and hammer tapping out the last grove in the girlfriends name. The old man looked up and smiled. He brought over the finished product. If you looked closely at the name, you could distinguish where the chiseled grove had been cut. Each gash was exactly the same in dimensions to the next.

The guy who had said the $0.25 a letter was too much now opined that we had screwed the poor old guy and the buyer should tip him. The buyer offered an additional dollar, which the old man declined.

We walked into another shop that sold all kinds of curios. We went about looking at variety of handmade items, jewelry, table decorations, and the like. I saw a black lacquer music box. I lifted the lid, and it began to play. It took me a second to recognize the tune; it was Sukiyaki. One of the other guys lifted the lid on a glass music box, and it too played Sukiyaki. We opened a third, and it played the same tune. The little old woman who ran the shop came over and closed the lid on the black lacquer music box. "You play only one." She admonished.

We took this as a challenge and wound up every music box in the store, opening them all. She promptly set about closing them. Soon we were in a whirl with us opening all the boxes and her closing them. "You GI numba ten bad. You not leave all running at same time. You break them." We were undeterred as we ran from music box to music box. She was just as determined to bring sanity back to her little shop. All the time yelling at us about being numba ten bad G.I.s.

At last we retreated into the open courtyard, her behind us yelling at us for being bad. We were all laughing. I was laughing so hard at the ridiculousness of it that my sides actually ached. As we started to walk back to the hospital, I suddenly realized I was laughing and my buddies were still dead and some others were still in harms way. I felt ashamed of myself. Where was my commitment to them? I sobered up, and we returned to the hospital.

That night I had a dream. It was one of the most vivid dreams I have ever experienced. I was walking in a jungle clearing. A fire was going in the center, and a guy was sitting on the log transfixed by the glowing embers. As I got closer, I could see it was Danny. I felt warmed by the sight of him, and I yelled out, "Hey!" He made no acknowledgement of me; in point of fact, he turned slightly away from me. I said hey again. Still there was no acknowledgement from him. "What's your problem man?"

"No, the question is what's yours?"

"What do you mean?"

"I mean, you had a pretty good time in that old woman's shop this afternoon, didn't you?"

"Oh that. We didn't harm anything; we were just having a little fun. We didn't break anything."

"Not the point, dummy."

"Well what is the point?" I asked

"The point is you were having a good time, and then you turned all dark and angry afterward. What'd you do that for?" There was pain in his eyes and a still sober countenance to him.

"You guys are dead. Some of the others are still in jeopardy. And I'm laughing like an idiot. That's not right."

"Why?"

"Cause you're dead!"

"Yeah, and you aren't, and that's the deal. I'm dead, and there's nothing you can do about. You're alive and good for you. So live."

"But I can't just go around like nothing happened. I have to honor you guys and your loss."

"So you're going to do it by wading into deep misery for the rest of your life? Is that what you would want me to do if the roles were reversed? You'd want me to be a miserable wretch hating everyone and everything?"

"Hey no, man, I wouldn't want that. I'd want you to go back and do all the things you had planned with Russell."

"So you're better than me? You'd want good for me, but I want nothing but an empty life for you?"

I was trapped in the illogic of it. Of all the guys I had gone over with, he was the one I had found the strongest base with. He was the one who seemed to have the best handle on life and what it was to like, and he was the one who had died. I was the screw-up, and I was going home, and now he was telling me I thought I was better than him. "No way. I would never think that. I know that's not true."

"Then wish for yourself what you would have wished for me if I was going home. Live and love and enjoy all of it. I ain't asking you to forget me or the others. I am asking you to not let it bog you down. Don't crawl into a cocoon of hate and misery. Think of us every once in awhile. But, live. Really live and enjoy every minute because you doing for us too."

I watched the fire and absorbed the heat. Danny talked about some of the things that had happened during our time together. I laughed, and so did he. As the fire ebbed into glowing coals, I found a comfort I hadn't had for quite a while, and I found myself making a promise: I wouldn't allow myself to become bitter and withered. I would try to live with as much joy as I could bring to each day. I would not let my friends down.

The next morning I became the thing I disliked most since I had been in the army: a mooch. I went around to everyone and tried to gather up enough money to go back and buy that black lacquer music box. Sukiyaki was now a hymn that meant life and love and all things good, and I intended to have that relic to tie me to it. I managed to gather enough and was about to head out when a nurse came in. I had to get my things together; I was taking the next evac flight, and I didn't have any time. I tried to persuade her that I'd be back in thirty minutes. I was advised I didn't have that much time.

I found myself on the tarmac in a fight with the loadmaster of the C-141 that was my transport home. I had a beautiful Fu Manchu mustache that he insisted on being trimmed back. My first attempt: I simply moved the ends up to even with the lower lip. This was met with rejection. My second trip, I cut the thing back to just at the ends of my mouth. This was also met with rejection. I finally realized that my upper lip would have to go home naked. As I emerged from the bathroom, I was met with a smile. "Now that's within regulation!" was all he said.

The huge transport took off from Japan and in a long arch flew to Anchorage, Alaska. There a bunch of Air Force wives met the plane. We were served coffee and donuts, and they chatted us all up. To a person they all thanked us for our service. I found that a very nice thing to hear. I walked out away from the plane and looked out across the tarmac. Huge piles of snow stood like guardians. Another C-141 parked near us had only its tall tail to mark its location. Alaska in April and the snow is that deep. It was going to take a lot of heat to melt all that. I found myself wishing I had the time to go back and buy the music box.

I knew that the world, for me, was going to be different because of that little box and the simple joy it had brought me.

We were soon back on the plane and headed back. Back to the world. Back to all that meant anything. I was alive in the factual sense and in the spiritual sense. I would try to fulfill my promise to Danny. We landed at Glenview Naval Air Station. We were off loaded and taken to the Naval Hospital by bus. On the way we passed a high school that was just letting out. Everyone stared out at the kids in their cars and walking along the street. I know I for one was wondering, had I ever been that young and innocent?

I was put on a ward with men suffering from various injuries. One guy had been badly injured by an explosion when he was working on the forty millimeter cannon on a Cobra. Another had lost his leg just above the knee when a friend stepped on a mine. As he said, he got better of it than his friend who was killed by the explosion. The first few days, I went through a battery of tests. I met a surgeon who would be working on me. The feeling in my hand and arm were gradually returning. I could feel all of my thumb and index finger. My middle finger was numb through the middle, and the ring and little fingers were still completely numb. The forearm along the ulna was still numb, but the inside was beginning to tingle like a foot that had gone to sleep.

That weekend my mom, sisters, and fiancé came to pick me up for the weekend. It was surreal. Less than a month past, I had been on a berm firing a machine gun into the tree line, and now I was riding in a car with my family. We went to Burger King and had a lunch that I had missed for the last eight months. My mom asked what I wanted to do, and I told her what I was doing that moment.

I spent the next four months going through surgeries and then doing staff work in various parts of the hospital. On weekends I spent time with my family and helped plan my wedding. After the surgery to repair the damaged nerves in my hand, an army lieutenant came through and gave me a box with my Purple Heart. I was told it was being awarded by a grateful nation. I got applause from the guys in the ward; most of them had gotten their Purple Hearts long before.

In August I got married. It was as natural a thing as eating or reading the morning paper. I had closed the door on Viet Nam, and I chose to seal off that event. I had a promise to keep to Danny, and dwelling on what had occurred in Viet Nam seemed, to me, counterproductive. After the wedding, I went to see if I could get a leave to go on my honeymoon. After being shuttled from office to office, I was sent down to Fort Sheridan. There I finally found a Specialist 4 who said I couldn't get a leave because there was no commanding officer for me. He told me I should go home and await orders. I only had thirty days on my enlistment, so I had no idea what orders I might be getting.

I was home until November 13th. I received a letter from the Department of the Army along with my DD-214. I was formally discharged from the army. The journey was complete. A line in the discharge read that I had been retained in the army an extra one month and twenty-four days for the convenience of the military, whatever that meant.

I was now enrolled in community college. I went to the VA and got set up for assistance with my school expenses. Because of my wounds, I would be getting an additional stipend. My wife was in her last year of nursing school so soon we would be set. I was now a work in progress with some successes and some failures. But that's the way of things when it's a work in progress.

I watched my country turn its back on our ally and throw away the sacrifice my friends had made, but I didn't want to let any of that drag me back down into the darkness. It would be all or none. So I chose none.

Epilogue

I spent a lot of time as a youth with my grandparents. My grandma had a phrase she used a lot: "I know." That phrase had three meanings depending on the circumstances. When I was hurt or sick, she would take me on her lap and rock and hum, and every once in a while she would murmur, "I know." She gave me the sense that she understood my suffering, and somehow that would abate my ills. There was warmth in that statement that shielded me during the worst of my woes.

When I was into mischief, she would advise me with a stern look and the comment "I know!" as if she had some detector that monitored my activities. I hated hearing those words. No matter how clever I thought myself, she would affirm her knowledge of my misdeeds. The fear of hearing those two words in that circumstance kept me out of trouble more than once, but not always. There was always that urge to tempt fate even with the specter of confrontation and the words "I know."

The last way she used the phrase was to express knowledge. That knowledge could be read, told to her, or wisdom divined from life. When she expressed knowledge on a subject with an "I know!" all were loath to contradict it. More worldly individuals might smugly disdain her view, but they weren't about to contradict Grandma's viewpoint.

I now use that phrase in the same three ways.

To all of you who served and suffered in that war, and it was a war, I say I know. I cannot ease your pain, and I won't pretend that what I went through has any comparison to your experience. We all had our moments, and we all faced them down to the best of our ability. Some of us fared better that others. A lot of that had to do with where we came from in the world and how we were received when we returned to that world. Our communities must hold some of that burden, and they will need to look inward and evaluate their attitudes. For me all I can say is, "I know!"

To the men, "the best and the brightest," who were in our national government I say, "I know." Politicians are soulless mirrors, a perfect reflection of the individual standing before them. Their core values are no deeper than the thin veneer behind the glass that allows them to seamlessly reflect the opinions of those who front them. You allowed the sacrifice of all those men and women to be eroded through your cowardice and duplicity. The many who fought, suffered, and died at your direction expected their effort to be honored by you keeping your word. What a foolish expectation. You were, after all, too smart to allow honor to be entered into the equation. There were elections to be won and lies to be told.

When the Paris Peace Accords were signed, promises were made to assure that those accords would be honored. What had gotten the North to the table was the continued sacrifices of the soldiers, sailors, airmen, and Marines who continued to fight and bleed in the fields of Viet Nam. I know that all of you who agreed to the accords and committed the United States to defend the settlement with financial and military support were lying. Unfortunately, so did the North Vietnamese. They may have had a small fear that the US would honor its commitment, but it was a small fear, and certainly not great enough to keep them from violating the accords before the ink was dry. The poor South Vietnamese government signed off on the agreement when they were assured of our financial and military support. But when it came nut cuttin' time, our purse strings drew tight, and you all pooh-poohed the South Vietnamese allegations of cheating by the North.

Your interests lay with your taste for political payback against a shamed president and his replacement. There was blame to be fixed and elections to be won. Honor be damned.

Lastly I say, "I know!"

I have a certain knowledge that we won that war. In every military confrontation, the men with boots on the ground prevailed. There was no circumstance where the units engaged did not prevail. They sacrificed life and limb to pursue and destroy an enemy, for the capturing of territory that planners would quickly abandon. They overcame the ridiculous limitations of a war conducted from Washington, D.C. by those whose only interest was the polling numbers and press coverage. In addition to the efforts, through combat, to defeat a relentless, remorseless, sometimes ruthless enemy, the men and women who served set about the task of winning over the population of peasants caught in the middle of that struggle.

We won, and that victory was spat upon. And even so WE WON that f***ing war.

To understand my enlistment in the army, you have to know something of the privileged life I was born into. Milford is a small farm community eighty miles south of Chicago straddling US Highway 1, the Dixie Highway. At its acme, when I was born, it was a thriving farm community typical of the Midwest in the mid-twentieth century.

For commerce there was a sale barn, Decker's, where all of the local farmers brought their livestock to sell. Buyers from a number of meat processing companies would sit and bid on cattle, hogs, and sheep. Farmers would compete with the professionals looking to acquire livestock that they might turn out to pasture during the summer and later bring back in to sell, having fattened them up during the long summer. A canning factory processed sweet corn, asparagus, and tomatoes that local farmers raised. In the late summer, the heavy smell of rotting cornhusks blanketed the whole town. Tractors trailing two and three wagons filled with sweet corn would fill the roads heading in to Milford. All life in Milford took on the pace of these wheeled vessels. My grandfather was a supervisor at the factory at one point in time, supplementing the family income. Several of my aunts and uncles worked there in the summers. One uncle, Richard, was an over-the-road trucker hauling the canned vegetables all over the country for much of his adult life.

There were two automobile dealers, for Ford and for Chevrolet. For retail there was a Ben Franklin five & ten store that carried everything from millinery to bubble gum cards. For groceries there was an A&P grocery store. Behind the Ford dealer, there was a lumber yard/hardware store. Everything a farmer could need sat in neat rows of bins that stood five feet high on both sides of narrow wooden aisles.

Faith was the bulwark of the community. Half a dozen churches of various Christian denominations met the religious needs of the people. No one, that I can remember, was a Bible thumping revivalist. I cannot remember anyone wearing their faith on their sleeve, but all had a sustaining belief in God. It was so strong that none had the need to argue it or question the existence of God. The only comment I ever heard regarding the various denominations was when my grandmother once commented that the thing about being a Catholic was that it took strong knees.

There was a small community bank where nearly everyone had a checking account and most everyone, when occasion required, borrowed money. The key there is required. No one borrowed for a new TV or a trip to Vegas. When there was seed to buy or fertilized, the required amount could be gotten. Those well known in the community could be trusted, and those who were trusted would die before betraying that trust. A man's word was everything.

Rounding out the community were two taverns where locals could have a beer and unwind from the days labors. Freedom was a given. No one thought about any of their activities in terms of being allowed by law.

My grandparents, Harold and Maude Hartman, had a small farm north and west of the town. One hundred sixty acres provided everything that my grandparents needed to raise their family of seven children. I lived there for the first three years of my life, and later I spent most of my summers there. This was the life that informed me. The love of country was imbued through the actions and conversations of my grandparents. The town was festooned with red, white, and blue in July, and there was no rush to take the bunting down. In the center of town, there was a memorial to all of the men from Milford who had served in the great conflicts of the twentieth century. Special designation was given to those who had been wounded and those who had died defending the country. I never imagined that someday I would see my name on the memorial. This was what I fought for. This is what America was to me, and this is what made it worth fighting for.